IoT Applications in Geotechnical Engineering

This book offers a comprehensive overview of how the Internet of Things (IoT) is reshaping the management of soil, geotechnical engineering, and environmental observations.

It explores extensive bibliographic research, examples, and case studies to provide nuanced information on improving the soil, evaluating its structural health, and forecasting geological disasters using IoT. Furthermore, it equips the reader with the necessary tools to apply IoT solutions to farming processes and increase the effectiveness of construction. By delving into the new developments within IoT sensors, geotechnical monitoring, and data analysis, the authors provide practical approaches to measuring soil humidity, assessing geological characteristics, and obtaining data through automation. It concludes with a call to pave the way for the wide-scale adoption of IoT for environmental supervision, construction activity, and precision farming.

This is a valuable resource for researchers, practitioners, engineers, and students of IoT-based soil management, geotechnical engineering, agriculture, construction, and environmental science.

IoT Applications in Geotechnical Engineering

Jagjit Singh Dhatterwal, Parveen Berwal,
Kuldeep Singh Kaswan, and Rajesh Goyal

CRC Press
Taylor & Francis Group
Boca Raton London New York Leiden

CRC Press is an imprint of the
Taylor & Francis Group, an **informa** business
A BALKEMA BOOK

Designed cover image: Getty Images

First published 2025
by CRC Press/Balkema
4 Park Square, Milton Park, Abingdon, Oxon, OX14 4RN

and by CRC Press/Balkema
2385 NW Executive Center Drive, Suite 320, Boca Raton FL 33431

CRC Press/Balkema is an imprint of the Taylor & Francis Group, an informa business

British Library Cataloguing-in-Publication Data
A catalogue record for this book is available from the British Library

ISBN: 9781032884080 (hbk)
ISBN: 9781032884097 (pbk)
ISBN: 9781003537625 (ebk)

DOI: 10.1201/9781003537625

Typeset in Times
by codeMantra

Contents

Author Biographies

Jagjit Singh Dhatterwal is an Associate Professor in the School of Computer Science and Artificial Intelligence, SR University, Hanumakonda Telangana, India. He is a Member of the Computer Science Teacher Association (CSTA), New York, USA, International Association of Engineers (IAENG), Hong Kong; and International Association of Computer Science and Information Technology (IACSIT), USA. He is also a Professional Member of the Association of Computing Machinery, USA, and IEEE; and a Life Member of the Computer Society of India. His areas of interest include artificial intelligence, BCI, cyborgs, and multi-agents technology. He has published several books, journal papers, and conference papers.

Parveen Berwal is a Professor in the Department of Civil Engineering at Galgotias College of Engineering and Technology, Greater Noida. He has an MTech and a PhD in Civil Engineering, with over 12 years of versatile experience in the field. He has 1[illegible] patents to his name and has contributed to numerous prestigious accreditations like NAAC, NBA, and NIRF. He has published more than 15 papers in several journals and international conferences. He is an active member of IRED (New York) and IAENG (Hong Kong). His areas of interest include highway construction, utilization of recycled aggregate, and use of waste material in civil engineering.

Kuldeep Singh Kaswan is a Professor in the School of Computer Science and Engineering, Galgotias University, Greater Noida, Uttar Pradesh, India. His contributions focus on BCI, cyborgs, and data science. He has 13 years of experience working with global universities like Amity University, Noida, India; Gautam Buddha University, Greater Noida; and PDM University, Bahadurgarh, India. He is also a senior member of IEEE; the Computer Science Teacher Association (CSTA), New York, USA the International Association of Engineers (IAENG), Hong Kong; and IACSIT, USA He has published several books, journal papers, and conference papers.

Rajesh Goyal is a Professor in the School of Construction and Dean at NICMAR University, Pune. He graduated in Civil Engineering from NIT Kurukshetra in 1998. He earned his PhD from IIT Roorkee in 2007. He is a Gold Medallist in MTech (Structures) from Kurukshetra University, Kurukshetra. His main areas of research are high-rise construction, construction management, new construction materials, and wind/tornado effects on structures. He conducted research in Germany, Japan, and the Czech Republic as a Postdoctoral Fellow under the prestigious DAAD Fellowship of Germany, MEXT Fellowship of Japan, and the Academy of Sciences Fellowship of the Czech Republic. Currently, he is a collaborative researcher in the Wind Engineering Research Centre, Tokyo Polytechnic University, Japan. He is also a visiting faculty of Selinus University, Italy. He has more than 50 research papers to his credit and has published in national and international conferences and journals.

Relevance of IoT for Soil Superiority

1

1.1 INTRODUCTION TO SOIL SUPERIORITY IN GEOTECHNICAL ENGINEERING

In geotechnical engineering, soil superiority appears to be a fundamental principle that defines safety, stability, and sustainability in any infrastructure project. The soil must possess qualities that guarantee its stability and reliability throughout the existence of buildings. This requires soil with optimal physical, chemical, and mechanical characteristics to support structures such as buildings, roads, and bridges. This chapter considers how new IoT technologies are enabling the eternal and safe preservation of soil superiority by engineers and how they have revolutionized the detection and prevention of problems (Behzadipour et al., 2023).

1.1.1 Defining Soil Superiority

Several factors must be considered to effectively define and determine soil superiority for construction. These include the type of soil, unit weight, void ratio, degree of saturation, and cohesiveness. In Ogbo, engineers use measurements of these factors to assess the soil's load-bearing capacity under structures. If not analysed properly, issues such as erosion, landslides, or subsidence can arise, leading to repair expenses and fatalities in extreme cases. These risks are mitigated by prioritizing soil superiority, enabling engineers to anticipate potential problems and adjust procedures as needed (Rohani et al., 2018).

DOI: 10.1201/9781003537625-1

1.1.2 Importance of Soil Superiority in Construction and Infrastructure

Soil superiority is a fundamental consideration in construction and development. For structures like skyscrapers, highways, or dams, the quality of the subsoil is just as important as the building materials. Poor soil quality can result in cracking, settlement or complete collapse. Traditionally, soil assessments were one-time processes involving sample collection and laboratory testing, which offered limited information. However IoT-enabling technologies now provide ongoing and real-time assessments of soil properties (Amraei et al., 2017).

In addition to stability, soil superiority affects other project characteristics, such as environmental sustainability and resource use. Favourable soil properties can improve water infiltration and drainage, minimizing soil erosion and reducing the need for extensive water control measures. Moreover, higher soil density can reduce the need for cement and aggregate, thereby cutting costs. Modern infrastructure requires the evaluation of soil conditions to assess natural hazards such as earthquakes, floods, or landslides in project areas. IoT technologies enable continuous data collection, allowing engineers to make informed decisions about soil treatment and reinforcement, ensuring the soil remains superior for structural support throughout the structure's lifespan (Selvanarayanan et al., 2024).

Geotechnical engineering has advanced with IoT technologies, improving construction efficiency and ensuring sustainability. IoT devices that monitor soil conditions help prevent issues like compaction, moisture evaporation, or subsidence, reducing the risk of structural failures and making infrastructure safer and more resilient in the long term (Behzadipour et al., 2023).

1.1.3 IoT Technologies for Soil Superiority

Considering the trends in geotechnical engineering, the application of IoT technologies for soil superiority reasons is becoming quite apparent. Whenever infrastructure requirements are being formed, both the quality and stability of soil are crucial to the safety of constructing structures. IoT brings forth high-end equipment that acts as a monitor of the real-time nature of the soil characteristics to minimize risks and improve decision-making. Application of these technologies in the analysis of soil enables engineers to track soil behaviour, hence producing better results in the short run as well as in the long run.

1.2 ROLE OF IoT IN MONITORING SOIL PROPERTIES

With the help of IoT, the concept of geotechnical engineering has become more effective in identifying the properties of soil (Table 1.1). More specifically, soil analysis was executed as a conventional method which required a significant period of time. Today, IoT has emerged as a boon to this sector as it offers engineers real-time monitoring of the health of the soil and its quality. IoT devices automatically collect and stream data continuously on various aspects of the soil, some of which are difficult to measure or assess.

TABLE 1.1 Role of IoT in monitoring soil properties

ASPECT	*DESCRIPTION*
Real-time data collection	IoT devices equipped with sensors collect real-time data on soil moisture, temperature, pH, and nutrient levels.
Precision agriculture	IoT systems enable farmers to monitor soil conditions and adjust water, fertilizer, and pesticide use for efficient farming.
Soil moisture monitoring	IoT sensors detect soil moisture levels, optimizing irrigation schedules to prevent overwatering or underwatering.
Temperature monitoring	IoT-enabled sensors track soil temperature, ensuring ideal conditions for seed germination and crop growth.
Nutrient level monitoring	IoT sensors measure soil nutrients (e.g., nitrogen, phosphorus, and potassium), guiding fertilization decisions.
pH level monitoring	IoT devices track soil pH, ensuring optimal acidity/alkalinity for specific crops.
Data-driven decision support	IoT systems provide actionable insights through data analysis, improving soil health and crop yield.
Automated irrigation systems	Integrated IoT sensors control irrigation systems, ensuring efficient water usage based on soil moisture data.
Remote monitoring	IoT allows remote monitoring of soil conditions, reducing the need for manual field inspections.
Early detection of soil degradation	IoT helps identify early signs of soil erosion, compaction, and degradation, enabling timely intervention.

(*Continued*)

TABLE 1.1 (*Continued*) Role of IoT in monitoring soil properties

ASPECT	*DESCRIPTION*
Environmental impact reduction	By optimizing resource use (water, fertilizers), IoT minimizes environmental impact and promotes sustainable agriculture.
Geospatial mapping of soil conditions	IoT devices can be used with GIS for mapping soil variability across large fields, enabling zone-specifi management.
Integration with weather forecasting	IoT integrates with weather data to predict soil condition changes, enabling proactive managemer strategies.
Soil health monitoring over time	IoT systems track soil health changes over time, helping farmers implement long-term soil conservation practices.
Cost efficiency and resource optimization	IoT reduces labour and resource costs by automatin soil monitoring and management processes.

Among all the features that IoT extends to monitoring the soil, it is more crucial t determine the moisture content of the soil accurately. Water content in soil is an esser tial component of construction, farming, and environmental conservation and form part of construction material and crops among other uses. Humidity sensors can detec even slight shifts in moisture content and relay the information to cloud host platform for evaluation. This real-time monitoring also guarantees that factors like drought stres or certain levels of water saturation are well detected and appropriate corrective mea sures taken (Rajendran et al., 2021).

Among the other soil characteristics, IoT also measures density. As for the geotech nical subfield, the density of soil plays an essential role in determining the stability c foundations and the resisting strength of structures. Sensors used for compaction of so through the use of IoT assist in the construction of roads, foundations, and other larg projects. If engineers periodically take density measurements, they can make decision based on them and minimize excessive compaction forces for maximum duration an safety (Babu et al., 2024).

1.2.1 Impact of IoT on Soil Moisture and Hydrology

Due to the numerous changes that IoT brings to the civil world, the changes to geotech nology and hydrology fall within the sphere of geotechnical engineering. In general, i can be concluded that IoT technologies have increased the degree of accuracy regardin the identification of moisture levels in the soil and offered information that is difficul to obtain in the context of the past. For instance, IoT enables the monitoring of th moisture level across fields in agriculture or construction sectors by installing device such as sensors across the field. The real-time information detailed above improves th likelihood of ascertaining the state of the soil health and determining the frequency of watering and utilization of the land.

1.2.2 Impact of IoT in Moisture Monitoring

One of the other significant upgrades is enhancing the efficiency of irrigation systems using IoT's help. Other conventional methods of watering are either overirrigation or underirrigation whereby they are both ineffective and can potentially harm a plant's growth. With regard to this challenge, IoT-enabled irrigation systems avail information relating to the moisture content of the soil among other factors. This information is utilized by smart irrigation controllers to adjust watering schedules and therefore water crops appropriately. Besides having this not detrimental effect on our water resource, it also helps in raising the yields and reducing the expenses of the farmers.

However, apart from irrigation, IoT applications have revolutionized groundwater utilization and control. Well water also forms part of the water mix as it requires timely and correct information in terms of its use to avoid depletion or pollution. Sensors to be installed in the well and well water recharged areas have the ability to determine the flow and quality of water on a real-time basis. It assists in the management of groundwater resources, problem identification at an early stage, and making decisions on the utilization and conservation of water resources.

1.2.3 IoT Sensors for Soil Quality Monitoring

One of the unique breakthroughs in this area is the use of IoT sensors to measure the quality of the soil. These sensors can accurately determine important product characteristics, including the level of moisture, temperature, density, and nutrient content. Used over a large area, these systems allow the collection of data that provides geotechnical engineers with extensive information about the state of the soil in Figure 1.1.

This constant monitoring helps provide remedies before they become emergencies within the area; for example, soil erosion, landslides, or even unfavourable situations for construction.

$$\theta_v = \frac{C_{\text{measured}} - C_{\text{dry}}}{C_{\text{wet}} - C_{\text{dry}}}$$

where

θ_v is the volumetric water content (soil moisture) in the soil.
C_{measured} is the capacitance measured by the sensor.
C_{dry} is the capacitance of the sensor in completely dry soil.
C_{wet} is the capacitance of the sensor in fully saturated soil.

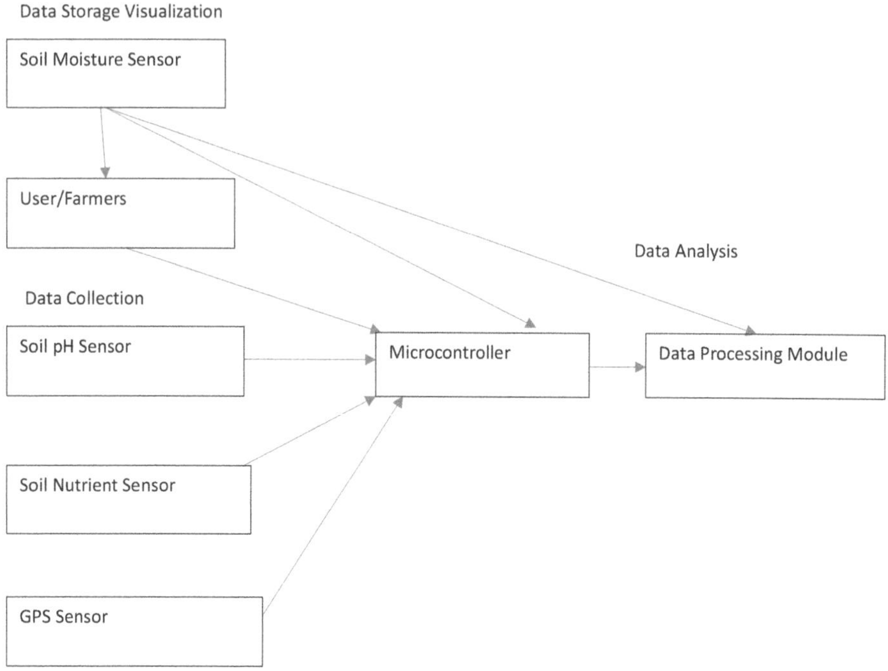

FIGURE 1.1 IoT sensors for soil quality monitoring

1.3 WIRELESS SENSOR NETWORKS (WSNs) IN SOIL ANALYSIS

Wireless sensor networks (WSNs) have a critical role in the IoT system for soil analysis. Such networks include the use of many sensors connected to each other in a network and conveying soil information to a control system through wireless signals. Compared to wired techniques, WSNs are convenient and can cover hard-to-wire locations or expansive premises. Such technology enables the collection of mass data from various areas, increasing the accuracy of assessments on soil.

Data storage and analysis relying on the cloud, combined with real-time analysis, add even more potential to IoT in geotechnical engineering. After it is collected from smart devices, data from IoT sensors are sent to the clouds where they are stored and processed. Cloud infrastructure delivers extensive scalability; this means that soil data can be stored and used when necessary. Furthermore, real-time analysis enables engineers to arrive at interpretations of the data much faster to adapt to changes in the soil.

$$P_{\text{total}} = \sum_{i=1}^{N} \left[P_{\text{sensor}}(i) + P_{\text{comm}}(i) + P_{\text{comp}}(i) + P_{\text{idle}}(i) \right] + P_{\text{gateway}}$$

$$+ \sum_{j=1}^{M} \left[P_{\text{data\_agg}}(j) + P_{\text{transonit}}(j) \right] - P_{\text{energy\_harvested}}(t)$$

where

P_{total} is the total power consumption of the WSN in soil analysis.

N is the number of sensor nodes in the network.

$P_{\text{sensor}}(i)$ is the power consumption of the ith sensor node for sensing soil parameters (e.g., moisture and temperature).

$P_{\text{comm}}(i)$ is the power used by the ith sensor node for communication (sending/receiving data).

$P_{\text{comp}}(i)$ is the power used by the ith sensor node for computation (processing data).

$P_{\text{idle}}(i)$ is the power consumed by the ith sensor node in idle mode.

P_{gateway} is the power consumption of the gateway node that aggregates data from multiple sensors.

M is the number of data aggregation points in the network.

$P_{\text{data\_agg}}(j)$ is the power used by the jth data aggregation point for aggregating soil data.

$P_{\text{transmit}}(j)$ is the power consumed by the jth aggregation point to transmit aggregated data to the gateway or base station.

$P_{\text{energy\_harvested}}(t)$ is the energy harvested over time t by the WSN (e.g., through solar panels).

1.4 CLOUD-BASED DATA STORAGE AND REAL-TIME ANALYSIS

The use of IoT technologies does not end with the collection and storage of data. Such assessment enables feedback and adjustments on the field in real-time continuation. For instance, if the ground of a construction site is not compacted enough in the required fashion, then IoT devices can inform engineers immediately and such a problem can grow into a major one later. This makes the management of the soil more dynamic in that the interventions made reflect only the most up-to-date knowledge.

IoT also incorporates cloud computing and wireless networks to enhance the collaboration of different geotechnical teams from various stations. Soil information is available for engineers for review, analysis, and co-ordination of plans to sustain soil supremacy from a distance. This kind of integration not only facilitates faster decision-making but also encourages result-oriented and integrated approach to managing the soil conditions for different projects irrespective of the region involved.

1.5 IoT-DRIVEN SOIL HEALTH MONITORING SYSTEMS

1.5.1 Components of IoT Soil Monitoring Systems

In the sphere of geological engineering, the IoT-based soil health monitoring systems are a technological breakthrough. These systems entail a web of associated sensors that constantly monitor the environmental states of the soil. Such components range from moisture sensors, temperature sensors, and soil compaction metres all with the aim of giving an overall view of the soil's state. Real-time data from such sensors is then fed to a central database that is in contact with such wireless networks for immediate analysis and response.

1.5.2 Integration of Artificial Intelligence and Machine Learning in IoT Soil Systems

Furthermore, the use of artificial intelligence (AI) in interaction with soil calibration helps in obtaining consistent readings of the sensors as well as determining the correctness of the data collected. One of the ways is the adjustment of the sensor accuracy incrementally to ensure that the data being collected is accurate and the monitoring system is efficient as shown in Figure 1.2. In addition, the monitoring parameters within an IoT environment can be applied for better targeting by machine learning algorithms which can adjust the mentioned parameters based on prior data broken down from other healthcare facilities. This makes the system react to changes in the soil and other external conditions in a manner that should suit the environment. The potential use of these highly superior systems has practical purposes in many fields. For example, through connected technologies and smart devices, IoT in soil health monitoring can enhance farming practices by embracing aspects of precision farming. In construction and infrastructure projects, the ability to monitor the condition of the soil in real time has become efficient in the overall decision-making processes regarding structural stability and health.

1.5.3 Data Collection and Analytics for Soil Superiority

Big data processing and data acquisition are vital aspects of IoT soil monitoring systems. Engineers using advanced analytics are in a position to analyse large volumes of soil data and determine factors that they would not have noticed otherwise. This capability makes it possible to have good control over the management of the soil and also

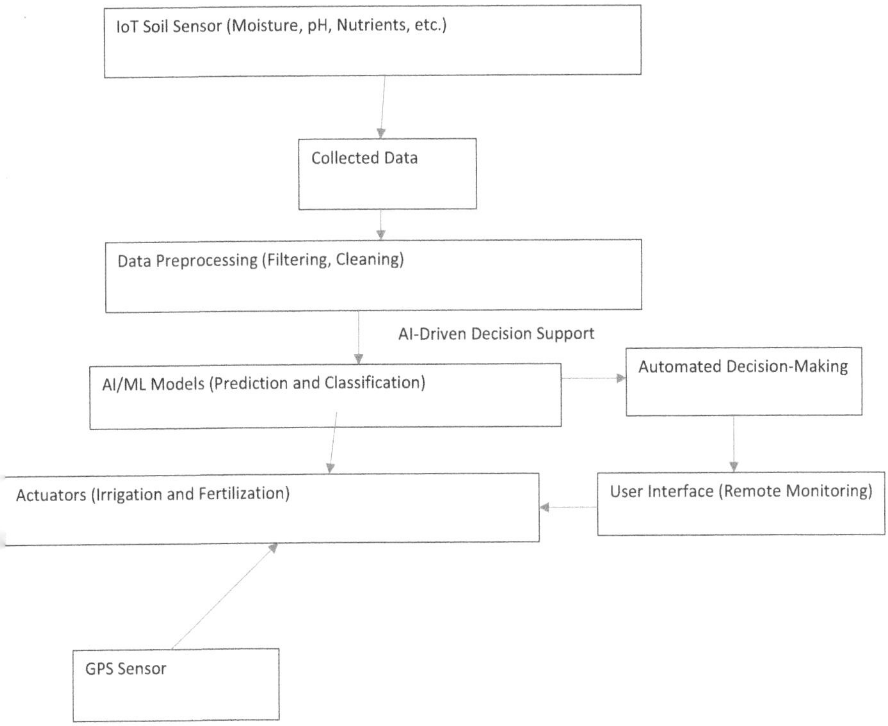

FIGURE 1.2 AI and machine learning in IoT soil systems

the prediction of adverse events before they occur. For example, current information on soil moisture can be used in times of spraying and planting to embrace efficiency in the use of water to gain more yields.

The integration of AI and machine learning into these IoT systems makes them even more effective. Data analytics can be used on historical and current data to identify the future state or behaviour of the soil and detect abnormality. This kind of predictability is, therefore, very useful in avoiding problems related to the soil and enhancing techniques in soil management. For instance, AI solutions can accurately predict possible occurrences such as landslides or soil erosion and prevent adverse incidences.

1.6 SOIL STABILITY AND IoT SENSORS

The stability of soil has been the foundation of geotechnical engineering, which plays the most crucial and decisive role in deciding the safety and durability of infrastructural structures. It is essential to take care of the loads on structures like roads, bridges, and buildings based on the stability of the soil that supports them. If stability is not well considered, there will be a high chance that structures will fail, resulting in cost repairs and, in extreme cases, total failure of structures. This makes soil stability a vit aspect of consideration and control among engineers in a bid to optimize for safety ar performance.

1.6.1 Importance of Stability in Geotechnical Engineering

The Internet of Things (IoT) helps geotechnical engineers to revolutionize the conce of soil stability. The IoT solutions provide updated information and data acquisition the engineers to determine the state of the soil in the best manner. IoT sensors may b used to measure the moisture content of soil, its compactness, and the stress it is unde The continuous feed of data to the central nervous system is especially useful in detec ing any signs of soil yielding before it becomes a serious problem that would requi urgent action. The latest development in this field is the use of WSNs in monitoring th stability of soil. Such networks consist of numerous sensors located at various parts (a given site, which relay information to a particular system, usually wirelessly (Plaza et al., 2020). This arrangement is suitable not only for obtaining all the necessary info mation but also for doing it quickly without the accumulation of large amounts of ne information or significant changes in the existing ones. By using this technology, eng neers can monitor the changes in the state of soil over time, which is essential in dete mining the effects of other factors, including weather or construction activities, on so stability (Rezk et al., 2021).

1.6.2 Sensors for Soil Moisture, Density, and Composition

Besides moisture and density, IoT sensors can determine the composition of the soil a well. This entails the identification of biological and inorganic matter, as well as trace of impurities. Soil characteristics concern the aggregate's ability to be used for a give purpose such as agricultural, construction, or any other specific use. Introducing Io sensors in geotechnical engineering allows geotechnical professionals to investigate th chemical and physical quality of the soil necessary to provide soils for the intended use as well as any contaminants or deficiencies (Satpathy et al., 2024).

WSNs of IoT devices are a crucial aspect of analysing soil concerning IoT systems They link different sensors installed on the soil in a way that they can share informatio with each other conveniently. Compared to checking the soil health in one location a a time, WSNs are able to scan a large area and determine the health of the soil simul taneously. This approach enhances the efficiency and quantity of information collecte concerning the soil, which in turn will lead to improvements in geotechnical project (Rajendran et al., 2021).

Cloud storage of data and parallel analysis, probably the most important components of any IoT-based system, cannot be excluded from the discussion about soil monitoring. The existing sensors generate tremendous data, and it has to be analysed and stored, which is where the concept of cloud computing comes into context. Due to the availability of the cloud, geotechnical engineers are able to view and analyse data pertaining to the soil from anywhere in the world whenever necessary for intervention. The cloud also allows the data from several sensors to be gathered and present an integrated view of the condition of the soil at different time intervals (Satpathy et al., 2024).

A. Soil Moisture:

Volumetric water content (θ)

$$\theta = \frac{V_w}{V_t}$$

where

- $C = \frac{\varepsilon_r \cdot \varepsilon_0 \cdot A}{d}$ θ is the volumetric water content
- V_w is the volume of water
- V_t is the total volume of soil sample

Dielectric constant relation (capacitance method)

$$C = \frac{\varepsilon_r \cdot \varepsilon_0 \cdot A}{d}$$

where

- C is the capacitance
- ε_r is the relative permittivity of the soil (related to moisture)
- ε_0 is the permittivity of free space
- A is the area of the plates
- d is the distance between the plates

T_{DR} (time domain reflectometry) for soil moisture

$$\theta = a + b \cdot T_{\mathrm{DR}} + c \cdot T_{\mathrm{DR}}^2$$

where

- θ is the volumetric water content
- T_{DR} is the TDR-measured dielectric constant
- a, b, c is the empirical coefficients

B. Soil density:

1. Bulk density (ρ_b)

$$\rho_b = \frac{M_s}{V_t}$$

where

- ρ_b is the bulk density of soil
- M_s is the mass of dry soil
- V_t is the total volume of the soil sample

$$\rho_b = \frac{M_s}{V_t}$$

2. Dry density (ρ_d)

$$\rho_d = \frac{M_s}{V_s}$$

where

- ρ_d is the dry density of soil
- V_s is the volume of solids in the soil sample

3. Porosity (n)

$$n = 1 - \frac{\rho_b}{\rho_s}$$

where

- n is the porosity
- ρ_b is the bulk density
- ρ_s is the particle density

C. Soil composition:

1. Grain size distribution (cumulative percentage passing)

$$P = \frac{W_p}{W_t} \times 100$$

where

- P is the cumulative percentage passing
- W_p is the weight of soil passing through a sieve
- W_t is the total weight of the soil sample

2. Soil pH (Hydrogen ion concentration)

$$\text{pH} = -\log_{10}\left[\text{H}^+\right]$$

where

- $\left[\text{H}^+\right]$ is the concentration of hydrogen ions in soil solution

3. Cation exchange capacity (CEC)

$$\text{CEC} = \frac{\sum (m_i \bullet v_i)}{M}$$

where

- m_i is the milliequivalents of exchangeable cations
- v_i is the valence of cation i
- M is the mass of soil sample

1.6.3 Early Detection of Contaminants via IoT Sensors

Integration of IoT technology in soil contamination detection also helps in enhancing regulatory compliance as well. Higher frequency and relevant data allow organizations to prove the implementation of environmental responsibility policies and maintain soil health. Besides, it assists the organizations in the fulfilment of their legal obligations and, at the same time, increases the level of public confidence in their activities.

However, it is not entirely perfect because there are matters that organizations and people should think twice about—one of them is a strong security requirement to avert an increasing number of cyber risks. Reliability and accuracy of IoT sensors are also very important because if something goes wrong with sensors or if there are inaccurate readings, they would completely decimate any soil monitoring initiatives.

1.6.4 IoT Solutions for Soil Stability Assessment

A number of examples and brief case studies are described below to show how IoT can forecast and mitigate soil-related problems. For instance, in regions at high risk of landslides, IoT sensors can detect changes in soil moisture content and movement and issue early warnings, enabling the evacuation of people and remedial actions to be taken (Hemdan & Al-Atroush, 2024). Likewise, in areas prone to erosion, the sensors can track erosion possibilities, allowing engineers to implement effective measures. Real-world applications, such as those mentioned, demonstrate how IoT may enhance soil stability, where the impact is tangible (Rezk et al., 2021). For example, in a large-scale infrastructure project, IoT sensors were employed to track soil conditions under a new highway construction. The data gathered helped engineers modify construction practices to ensure the proper support pressure for the soil and protect against environmental conditions. As a result, the highway became more stable, minimizing future maintenance problems (Kaswan et al., 2024).

1.7 CASE STUDIES: IoT FOR LANDSLIDE AND EROSION PREDICTION

In environmental monitoring, IoT integration is also crucial for assessing soil stability. Supplementary data on soil moisture, temperature, and other factors help engineers understand how environmental changes impact soil stability. Such information is essential for developing strategies that accommodate various scenarios, leading to more sustainable environments. However, there are challenges to consider such as sensor placement, data integration, and system design. Nevertheless, the potential to reduce risks associated with soil instability outweighs these considerations, making IoT a valuable tool in current geotechnical engineering practices (Plazas et al., 2020).

1.8 IoT SOLUTIONS FOR GROUNDWATER MANAGEMENT

The application of IoT in soil moisture and hydrology helps to improve the practice of sustainable land use. Finally, IoT's specificity and localization enable approaches to be implemented to conserve soil and water resources, thanks to the detailed data. For instance, accurate moisture levels help direct the application of cover crops or conservation tillage procedures that minimize soil erosion, as well as enhance water storage (Dhatterwal et al., 2024).

However, some issues remain in the use of IoT for Smart Manufacturing

(SM) and hydrology. Challenges like data integration, maintenance of sensors, and data accuracy require constant regulation. Nonetheless, as the technology grows, these issues are expected to be solved, hence improving the use of IoT in the management of soil and water resources.

1.9 IoT AND SOIL COMPACTION MONITORING

This chapter therefore postulates that the challenge of soil compaction is central to the stability and sustainability of infrastructure projects. Well-compacted soil provides good support for structures, roads, and pavements, which reduces settlement and other structural failures in the future. Work done essentially determines the strength and load-bearing capacity of the soil and serves as the foundation for civil engineering and construction (Batista et al., 2004).

1.9.1 Importance of Soil Compaction for Infrastructure

In the past, the amount of compaction was determined using trial pocket tests, which, although a reasonably accurate way of determining the state of compaction, offered a restricted view and took a comparatively long time to perform. The application of IoT technology in agriculture has significantly increased the prospect of qualitative evaluations for assessing compaction in soils. Today's IoT technologies include sensors and wireless networks that detect and report compacted soil density and moisture measurements, enabling engineers to evaluate compaction quality in real-time and assess the actual strength (Togneri et al., 2022). These tools assist in compacting the soil to the required standard and help align techniques as effectively as possible.

1.9.2 IoT Tools for Monitoring and Optimizing Soil Compaction

The usage of IoT in soil compaction monitoring is not just theoretical but based on actual examples. For instance, in road construction, IoT sensors have been used to assess the impact of rollers on the compaction of asphalt and subgrade. These sensors relay information regarding the soil environment to central systems, enabling engineers to review the data received and make real-time corrections to ensure roadbeds meet specifications for durability and functionality (Islam et al., 2023).

1.10 APPLICATIONS OF IoT IN SOIL CONTAMINATION DETECTION

The other advantage of using IoT systems is that they help identify contaminants in the early stages. Many of the traditional methods used to test the soil take a lot of time to complete and all that is confirmed is contamination when it is severe. However, IoT sensors do allow engineers, in addition to environmentalists, to know of a possible issue as soon as it takes place. This is a more aggressive approach that can lead to early action that is important for mitigating the impacts of other environmental effects and preventing adverse health effects (Togneri et al., 2022).

1.10.1 IoT Systems for Monitoring Soil Pollution

Some of the industrial samples have illustrated how IoT can be useful for regulating the purity of the soil. For example, in agricultural environments, IoT devices are utilized to detect such contaminants as toxic chemicals or metals that endanger crop safet Through the application of soil sensors on irrigation systems, farmers are in a positic to detect cases of pollution and take the necessary action before they affect the plant Likewise, in the industrial zones and regions, IoT technology assists in monitoring th ground situation where pollutants are produced and spread, and it avoids damagir other zones.

1.11 CASE STUDIES: INDUSTRIAL APPLICATIONS OF IoT FOR SOIL PURITY

Use cases explain how IoT is being implemented in the business world in differe fields. For example, there is information on how a mining company installed IoT ser sors to check the concentration of soil pollutants around the company's operation Real-time data enabled them to react when levels of pollutants went up above the rec ommended limit, hence minimizing the effects on the environment. Another exampl was one of the agricultural firms that applied IoT to track soil receptiveness to fertiliz and contamination by industrial effluents for healthy crop growth and soil enhance ment, respectively.

Therefore, in the future, there will be enhanced potential for IoT in the identif cation of soil contamination. Forthcoming improvements in sensor mechanisms an heightened big data analytics guarantee higher levels of accuracy in the monitorin mechanisms. They will continue to take a more important part in securing soil qualit and creating a better future for all generations.

1.12 IoT IN PREDICTIVE MAINTENANCE OF SOIL CONDITIONS

Discussed in the current contribution is the fact that IoT influences the geotechnica engineering field significantly, specifically by enabling accurate predictive maintenanc concerning the state of soils. Before it becomes problematic, there is a crucial predictiv role that models are assigned to perform regarding soil degradation. Through IoT sen sors, the models identify patterns in the health state of the soil and foresee challenges allowing preventive measures to be taken. Such a proactive approach enables the contrc measures to be applied on time in order to avoid digging expensive holes and conserv ing soil integrity in the long run.

1.12.1 Predictive Models for Soil Degradation

The IoT solutions regarding preventive actions are evolving higher in complexity. Various stimuli, including moisture, temperature, and nutrient contents in the soil, are measured by tiny devices incorporated in the soil. Real-time data is obtained from these sources and entered into advanced estimator models for analysis that can be used to formulate specific strategies for soil conservation. For instance, if a particular sensor measures a decrease in soil moisture, the system can in turn control irrigation, making the soil always conducive for plants.

Uniquely, the application of IoT in agricultural land management has been fruitful. The various examples and case studies have illustrated how IoT systems can improve the quality of the soils and their yields in a huge way. For instance, one of the detailed cases described the situation when IoT sensors helped a big farm to control the current state of soil and make corresponding changes in the fertilization plan. It also had positive effects in minimizing wastage and negative effects on the environment since scarce resources were applied properly.

1.12.2 IoT-Enabled Preventive Measures for Soil Health

Another of the presented case studies related to the application of IoT in controlling soil erosion. Specific instrumentation applied in the high-risk zones helped monitor the ground activity and rates of soil degradation. Such knowledge could be used to make decisions regarding the use of specific gadgets with objectives to control the rate of erosion and, thereby, conserve the valuable topsoil on the land concerned.

IoT systems have also been employed to quantify the salinity of soil, which might be a nuisance in arid regions. The farmers may change their irrigation practices to avoid the effects of soil salinization and maintain the fertility of the soil, while preserving crop productivity by analysing information collected from the salinity sensors.

Taking everything into account, IoT has a relevant contribution to the area of predictive maintenance of soil conditions. IoT enhances understanding of sustainable land management practices and proper maintenance of the soil through real-time monitoring and quick response. IoT will positively contribute to the enhancement of agricultural yields and effective soil control in future appliances through early management and conservation of the soil. This is especially the case as technology continues to grow.

1.13 SUSTAINABILITY AND SOIL SUPERIORITY THROUGH IOT

Speaking of interconnected sustainability and the superiority of soils by using IoT, it is quite obvious that this path paves the way to further sustainable handling of soils.

The adoption of IoT in the management of soil sustainability cannot be ignored since it provides real-time data that dictates management decisions. IoT systems provide an opportunity to track the conditions of the soil and the environment and identify the right practices for its maintenance and usage with the interference of people. The application of these techniques also prevents wastage of the resources used in managing the soil, hence making a difference.

1.13.1 Promoting Sustainable Soil Management

Speaking about the IoT technologies' applications, one should mention how it is useful in fields such as soil conservation and rehabilitation. Modern sensing technologies and remote monitoring enhance the understanding of the condition of the soil and the extent of its deterioration so downstream conservation efforts can be specific. For example, irrigation can enable determination of a precise pattern of soil erosion, nutrient depletion, and moisture content to apply the right conservation techniques. It not only has preventive implications for reducing additional deterioration of soil but also usefully contributes to the restoration of damaged terrains, promoting the strengthening of permanent functioning in the sphere of soil usage.

1.13.2 IoT for Soil Conservation and Rehabilitation

It also poses the environmental effects of the IoT in soil management. It enables better results on the soil by reducing the application of excess fertilizers and by advising and advancing data-driven soil management that has impacts on the environment. Besides, IoT systems provide suggestions regarding the use of water and how this valuable resource can be used most efficiently and effectively with reduced wastage and likelihood of polluting the water closest to the irrigation systems. The general impact of IoT on soils includes less concentration of chemicals which leads to minimum chemical runoff, efficient water utilization and quality improvement of the soil, and an overall balanced environment.

1.13.3 Environmental Impact of IoT in Soil Management

Therefore, as IoT technology continues to advance, its enhanced capabilities will further contribute to sustainability improvements and, consequently, the enhancement of soil quality. Because real-time information is available and more efficient decisions are made by IoT, long-term strategies for maintaining sustainable soil management which are viable and eco-friendly are promoted. The continuous and progressive advancement in IoT presents an opportunity to create only healthy soil, and innovative and intelligent concepts can help in supporting sustainability principles.

1.14 CHALLENGES AND SOLUTIONS FOR IMPLEMENTING IoT IN SOIL SUPERIORITY

There are certain problems that appear when applying IoT in soil superiority, and to optimize its applicability, these issues must be considered. The first technical and operational issue is the reliability and accuracy of sensors and systems involved in the process. There are diverse types of soils, and the sensors must be able to give sound results regarding the differences in the soil type. These types of sensors should be calibrated and maintained in a correct manner so that the corresponding data will not be incorrect. Furthermore, implementing IoT systems into soil management systems can be complex and may need changes or improvements in the existing practice, which may come with different costs and technical know-how.

Next is scalability: Every business needs to be able to scale as the demand grows; a business idea that does not have this ability will struggle to survive. As IoT systems transition to larger areas or if the soil management issues to be dealt with become more complicated, the ability of the technology to scale becomes critical. As part of these responsibilities, one must ensure a higher volume of data and the efficiency of the monitoring systems. Implementation may also raise issues of interface with other systems also, as most apply inputs, utilities, and manipulation of existing data management systems. These solutions must be easy to implement on large scale and furthermore be easily incorporated into existing structures and architecture, specifically for IoT applications for better soil management.

It is crucial to mention that security and data privacy are the most pressing issues in deploying IoT systems. Soil sensing data is mostly discreet and can significantly influence decision-making among various stakeholders. It therefore becomes necessary to make sure that this information is safe from such activities as hacking. Reducing these risks can be achieved by integrating strong cybersecurity measures such as data encryption and secure storage of data. Besides, transparency in data management and processing and adherence to the data protection guidelines are useful for building trust and overseeing the proper use of IoT technologies.

1.15 FUTURE TRENDS IN IoT FOR SOIL SUPERIORITY

Assuming the future of IoT in soil superiority and the factors that will define it, several trends and technological innovations can be outlined. In view of this, technologies in IoT are relatively new and are bound to improve in precision and capabilities of soil monitoring. Structural improvements in the mechatronic sensing system will improve the sensitivity and longevity of the sensors, enabling the assessment of additional soil properties with greater accuracy and precision. Derivatives like nano-sensors and better data analysis tools will enhance soil tests in the future, providing a better understanding of the state of the soils, and hence, better management is bound to improve.

Linking IoT with remote sensing and geographic information systems is a major step forward in soil management. Integrating IoT information with other remote sensing technologies, including satellite and aerial drones, will provide an improved understanding of the state of soil over larger fields. This integration led to real-time monitoring and analysis at a scale that could not be done before. This will be complemented by Geotechnical Geographic Information System (GGIS) through a graphical display of the data to capture the spatial dimensions of the soil map and trends to enhance the interpretation of the data through mapping and identifying soil variations that can support decision-making.

Smart soil management systems are yet to come, and the world is likely to witness more intelligent and automatic solutions for managing soils soon. These systems will utilize the IoT data paralleled with AI as well as the machine learning solution to make necessary and timely changes for efficient soil management. For instance, smart irrigation could use soil moisture data to control irrigation levels while predictive models could predict soil conditions and give advice on ways to avoid future problems. Apart from increasing efficiency, this approach also helps maintain proper soil usage.

1.16 CONCLUSION

Now sitting perfectly for IoT harmonization, geotechnical engineering is getting a brand-new approach to the ways of achieving soil superiority—with improved accuracy, efficiency, and control. Thus, we outline the concept of soil superiority and its importance for construction and infrastructure to understand how IoT technologies improve the quality and stability of the soil. IoT is composed of sensors that track moisture content, density, and soil composition; real-time data analysis helps engineers make proper decisions to improve performance and durability. Some of the emergent IoT technologies include WSNs and the cloud, which facilitate continuous collection and analysis of data and revolutionize soil health monitoring. Current advancements in the application of AI and machine learning mean that there are models of predictive maintenance to avoid issues with soil degradation before they are critical. Examples include landslide control, soil condensation for construction purposes, as well as the identification of contaminations, in practice. The effects of IoT on moisture and hydrology show that the control of irrigation systems and groundwater may be changed with the help of IoT technology and sustainability measures using IoT-based conservation. Despite challenges like scalability, technical glitches, and data protection, IoT has a very positive outlook towards soil superiority. While IoT and smart systems emerge and provide new opportunities for use in geotechnical engineering, the importance of IoT in future soil management will expand to new levels of utilization, contributing to its efficiency and sustainability.

REFERENCES

Amraei, S., Abdanan Mehdizadeh, S., & Salari, S. (2017). Broiler weight estimation based on machine vision and artificial neural network. *British Poultry Science*, *58*(2), 200–205.

Babu, S., Madhusudanan, S., Sathiyanarayanan, M., Mortka, M. Z., Szymański, J., & Rahul, R. (2024). Soil mapping for farming productivity: Internet of things (IoT) based sustainable agriculture. *Microsystem Technologies*, 1–16.

Batista, G. E., Prati, R. C., & Monard, M. C. (2004). A study of the behaviour of several methods for balancing machine learning training data. *ACM SIGKDD Explorations Newsletter*, *6*(1), 20–29.

Behzadipour, F., Ghaseminezhad, M., Abdanan Mehdizadeh, S., Taki, M., Khalili Moghadam, B., & Zare Bavani, M. R. (2023). Predictions of greenhouse soil moisture using artificial neural network and wireless network sensing. *Iranian Journal of Biosystems Engineering*, *53*(4), 341–356.

Dhatterwal, J. S., Kaswan, K. S., & Chithaluru, P. (2024). Agricultural cyber-physical systems: Evolution, basic, and fundamental concepts. In Seifedine Kadry, Vandana Sharma, Rajesh Kumar Dhanaraj, Rutvij H. Jhaveri, Gandhiya Vendhan (Eds.), *Agri 4.0 and the Future of Cyber-Physical Agricultural Systems* (pp. 19–35). Academic Press.

Hemdan, E. E. D., & Al-Atroush, M. E. (2024). An efficient IoT-based soil image recognition system using hybrid deep learning for smart geotechnical and geological engineering applications. *Multimedia Tools and Applications*, *83*, 66591–66612.

Islam, M. R., Oliullah, K., Kabir, M. M., Alom, M., & Mridha, M. F. (2023). Machine learning enabled IoT system for soil nutrients monitoring and crop recommendation. *Journal of Agriculture and Food Research*, *14*, 100880.

Kaswan, K. S., Dhatterwal, J. S., Batra, R., Balusamy, B., & Gangadevi, E. (2024). 3D Bioprinting technology optimization using machine learning. In E. Gangadevi, M. Lawanya Shri, Rajesh Kumar Dhanaraj and Balamurugan Balusamy (Eds.), *Computational Intelligence in Bioprinting* (pp. 303–321). Scrivener Publishing.

Plazas, J. E., Bimonte, S., de Vaulx, C., Schneider, M., Nguyen, Q. D., Chanet, J. P.,... Corrales, J. C. (2020). A conceptual data model and its automatic implementation for IoT-based business intelligence applications. *IEEE Internet of Things Journal*, *7*(10), 10719–10732.

Rajendran, S., Khalaf, O. I., Alotaibi, Y., & Alghamdi, S. (2021). MapReduce-based big data classification model using feature subset selection and hyperparameter tuned deep belief network. *Scientific Reports*, *11*(1), 24138.

Rezk, N. G., Hemdan, E. E. D., Attia, A. F., El-Sayed, A., & El-Rashidy, M. A. (2021). An efficient IoT based smart farming system using machine learning algorithms. *Multimedia Tools and Applications*, *80*, 773–797.

Rohani, A., Taki, M., & Abdollahpour, M. (2018). A novel soft computing model (Gaussian process regression with K-fold cross validation) for daily and monthly solar radiation forecasting (Part: I). *Renewable Energy*, *115*, 411–422.

Satpathy, S., Khalaf, O., Kumar Shukla, D., Chowdhary, M., & Algburi, S. (2024). A collective review of Terahertz technology integrated with a newly proposed split learning based algorithm for healthcare system. *International Journal of Computing and Digital Systems*, *15*(1), 1–9.

Selvanarayanan, R., Rajendran, S., Algburi, S., Ibrahim Khalaf, O., & Hamam, H. (2024). Empowering coffee farming using counterfactual recommendation based RNN driven IoT integrated soil quality command system. *Scientific Reports*, *14*(1), 6269.

Togneri, R., dos Santos, D. F., Camponogara, G., Nagano, H., Custódio, G., Prati, R.,... Kamienski, C. (2022). Soil moisture forecast for smart irrigation: The primetime for machine learning. *Expert Systems with Applications*, *207*, 117653.

2 Assessment of Soil Humidity Content Using Internet of Things and Conventional Systems

2.1 INTRODUCTION

In soil, humidity holds significant importance in geotechnical engineering as well a agricultural sciences, where determining the humidity level can exert a considerabl impact on the results of constructed facilities and cultivated plants. Some of the mos typical historic approaches to determining soil moisture include gravimetric measuring tensiometers, and gypsum blocks. While these techniques provide reliability, they com with inherent disadvantages such as inaccuracy, the inability to provide real-time data and difficulty of use (Pamula et al., 2022).

Internet of Things (IoT) technology introduces exciting prospects for how soi monitoring has been traditionally conducted. IoT-based systems offer data collectio and monitoring from a distance with greater accuracy than traditional methods. By integrating more precise sensor networks, reliable signal transmission techniques, and cloud computing for data storage and analysis, these systems are transforming soi

DOI: 10.1201/9781003537625-2

moisture evaluation. Capacitive sensors, resistive sensors, and time-domain reflectometry (TDR) are key technologies enabling this transformation (Krishnan et al., 2020).

A comparison between IoT systems and conventional techniques reveals numerous advantages: more precise results, immediate feedback, greater data accessibility, and higher energy efficiency. IoT has demonstrated its applicability in managing soil moisture for agricultural and geotechnical applications (Darsi et al., 2021).

However, challenges remain, including technical issues, financial constraints, sensor calibration, and data security. Despite these challenges, future advances in sensor technology, the use of artificial intelligence for predictive analysis, and even blockchain-based smart sensors are on the horizon (Ravilla et al., 2024). Smart construction and smart farming continue to integrate IoT, promising a bright future.

The gravimetric method is one of the oldest techniques. It involves taking soil samples, weighing them, drying them, and then weighing them again. Though accurate, this method is time-consuming, non-repeatable, and unsuitable for real-time monitoring (Pamula et al., 2022). While tensiometers and gypsum blocks provide single-point estimates of the soil matrix potential, their performance declines over time without frequent recalibration. These conventional systems are also impractical for large-scale agricultural or construction projects.

In future geotechnical engineering projects, the necessity of standardizing the size and complexity of conventional soil humidity measurement techniques becomes even more evident. Traditional methods rarely portray changes due to weather variations or shifts in irrigation regimes. Consequently, there has been an increasing need for sophisticated technological platforms, such as IoT, to monitor soil in real-time.

IoT-based systems have significantly shifted environmental monitoring by providing real-time, easily scalable, and highly accurate solutions. Networks of sensors distributed across soil capture moisture data at regular intervals and transmit it via communication networks to cloud-based monitoring systems (Krishnan et al., 2020). This information enables swift interventions, whether adjusting watering cycles in farming or addressing soil shifts in construction.

Various IoT sensors, including capacitive, resistive, and TDR sensors, are used to provide general moisture information over expansive areas. Data is transmitted using low-power radio technologies like LoRa, Zigbee, or NB-IoT, offering enhanced precision and minimizing efforts and time (Pamula et al., 2022).

In construction and agriculture, IoT-based real-time monitoring systems provide a competitive edge over traditional approaches by allowing stakeholders to act promptly based on changing conditions. IoT also contributes to sustainability by optimizing water use and reducing waste. However, like any technology, IoT systems face challenges, including high fabrication costs, sensor calibration issues, and data privacy concerns (Gopikumar et al., 2021). As the technology advances, improvements in sensor reliability and data security through blockchain are addressing these concerns.

2.2 CONVENTIONAL SYSTEMS FOR SOIL HUMIDITY MEASUREMENT

Soil humidity is critical in geotechnical engineering to determine the stability of construction projects and the productivity efficiency of agricultural land. Originally, engineers used other conventional systems to measure soil moisture.

2.2.1 Overview of Traditional Techniques

These methods have been proven over a period of time and provide only a fundamental level of knowledge about the water in the soils, which in turn plays a significant role with regard to the behaviour of the soils. However, as with any conventional systems, they present their own particular difficulties that have to be considered in order to put the benefits and limitations into perspective.

The gravimetric method is one of the most popular traditional methods used to determine soil moisture. This technique involves the collection of a soil sample that is weighed, after which it is dried using an oven and then weighed again to deduce the moisture by the differences in weight. While very precise, the gravimetric method is tiresome, time-consuming, and destructive in that actual soil samples have to be collected from the site.

$$\theta_g = \frac{m_w}{m_s}$$

$$\theta_v = \theta_g \cdot \frac{\rho_b}{\rho_w}$$

$$\Psi_m = -\frac{2\gamma \cos(\alpha)}{r}$$

$$S = \left(\frac{\Psi_m}{\Psi_0}\right)^n$$

$$\frac{d\theta}{dt} = -K(\theta)\left(\frac{d\Psi_m}{dz} + 1\right)$$

$$\Psi_{\text{tensio}} = -h_t \cdot \rho_w \cdot g$$

$$\Psi_{\text{gypsum}} = f\left(E_{\text{output}}\right)$$

$$K(\theta) = K_s \cdot \left(\frac{\theta - \theta_r}{\theta_s - \theta_r}\right)^\lambda$$

$$q = -K(\theta) \cdot \frac{d\Psi_m}{dz}$$

- θ_g is the gravimetric water content
- m_w is the mass of water
- m_s is the mass of dry soil
- θ_v is the volumetric water content
- ρ_b is the bulk density of soil
- ρ_w is the density of water
- Ψ_m is the soil matric potential
- γ is the surface tension of water
- α is the contact angle
- r is the pore radius
- S is the degree of saturation
- Ψ_0 is the reference matric potential
- n is the empirical fitting parameter
- $\frac{d\theta}{dt}$ is the rate of change of soil moisture
- $K(\theta)$ is the hydraulic conductivity as a function of θ
- $\frac{d\Psi_m}{dz}$ is the hydraulic gradient
- Ψ_{tensio} is the tensiometer-measured soil water potential
- h_t is the height of the water column in the tensiometer
- g is the gravitational acceleration
- Ψ_{gypsum} is the soil water potential from gypsum blocks
- E_{output} is the electrical signal from the gypsum block
- K_s is the saturated hydraulic conductivity
- θ_s is the saturation water content
- θ_r is the residual water content
- λ is the pore-size distribution index
- q is the water flux

Tensiometers and gypsum blocks can be mentioned as other conventional types of soil moisture sensors. Tensiometers reflect the useful tension or vacuum which the plant draws out of the soil. It gives real-time results and is less intrusive than the gravimetric approaches for evaluating temporal changes in soil water content. Gypsum blocks, in contrast, determine soil moisture concentration by measuring the resistivity of a block placed into the soil. However, such methods are more appropriate for continuous assessment and are usually less precise compared to the initial method, with sensitivity to soil type and weather conditions.

2.2.2 Advantages and Limitations of Conventional Systems

Traditional methods have their own advantages and disadvantages, and the table below outlines some of them. The gravimetric method, for example, is very accurate but cannot be used for real-time monitoring or monitoring on a larger scale. Tensiometers are helpful for monitoring measurements in real-time but are influenced by the soil salinity level and temperature. The costs of gypsum blocks are relatively low, and installation is simple. However, its accuracy declines in high salinity and blocks degrade with time.

Comparing the accuracies and reliability of these conventional systems made me understand that even though each has its advantages, none is 100% effective. The gravimetric method requires high accuracy, but it is not appropriate for dynamic or large-area measurements. Tensiometers and gypsum blocks are more applicable for real-time measurements; however, these techniques can lack accuracy and are greatly influenced by the variability of the soil. Moreover, some of these tools like tensiometers may need constant refilling and checkups. Additionally, gypsum blocks deteriorate, resulting in inaccurate measurements.

Even though these approaches present various drawbacks, they are still relevant as methods for traditional analysis, especially in areas where access to high technologies is still impossible. Their reliability in applications that are unilateral, simple, and small-scale makes them quite handy, but they are desirable in applications where precision and scalability form the high-end goals of the project.

2.2.3 Comparative Analysis of Accuracy and Reliability

On the other hand, new approaches which are based on IoT redefine some of the limitations of conventional systems. Mainstream servers are not capable of long-term, distant observation and therefore are more suited to much larger, intricate projects where instant information is crucial.

Nonetheless, such traditional approaches are within reach of most implementations since they involve low cost, especially compared to IoT systems. Nonetheless, as newer technologies are developed, introducing technical means is gradually becoming obligatory in industries that use tools with great accuracy.

2.2.4 Advantages of IoT-Based Systems over Conventional Methods

Conventional methods of testing and monitoring the state of soil cannot be compared to IoT-based systems, which are capable of monitoring soil constantly and in real-time. Traditional approaches, such as tensiometers or gravimetric methods, while accurate in small samples and controlled conditions, are often manual and provide only intermittent information for large fields, with results varying widely (Rajak et al., 2023). IoT systems, however, provide constant observation, reducing the time needed to detect shifts in soil status. For applications like agriculture and geotechnical engineering, real-time feedback is essential for efficient decision-making (Morchid et al., 2024).

The IoT soil monitoring systems are anchored on sensors, as they are responsible for collecting data. These include capacitive sensors, resistive sensors, and time-domain eflectometry (TDR) sensors, which are capable of measuring even small variations in oil moisture. In some cases, specific sensors are chosen to suit the application, ensuring ccurate measurements. Many of these sensors are inexpensive, consume low power, nd are designed for long-term durability under varying external environments (Rajak t al., 2023).

2.2.5 Comparison between IoT-Based and Conventional Systems

Several significant differences in technologies are revealed when comparing IoT-based soil humidity measurement systems with traditional approaches in terms of accuracy, precision, real-time monitoring, data accessibility, and environmental friendliness in Table 2.1. Each approach has its own benefits, yet IoT systems are rapidly favoured because of better and updated technologies integrated within them (Sinha & Dhanalakshmi, 2022).

TABLE 2.1 Comparing IoT-based soil humidity measurement systems with conventional systems

CATEGORY	*IOT-BASED SYSTEMS*	*CONVENTIONAL SYSTEMS*
Accuracy and precision	Provides high accuracy and precision using advanced sensors (e.g., capacitive, resistive, TDR) (Sinha & Dhanalakshmi, 2022). Real-time data refinement is available	Conventional methods like gravimetric analysis and tensiometers provide accurate measurements but are less suitable for large-scale or long-term monitoring (Boursianis et al., 2022)
Response time	Offers real-time monitoring with instant feedback, enabling quick responses to soil moisture changes (Dutta et al., 2024).	Data is often collected manually at regular intervals, leading to delays in identifying changes (Giannakidou et al., 2024)
Data availability and access	Data is accessible remotely via cloud technologies. Real-time data can be accessed from any location through a user interface (Feng, 2011)	It requires field visits for data collection, limited by sampling points and offers only static data snapshots (Boursianis et al., 2022)
Energy efficiency	Designed for low power consumption. IoT sensors often run on low-energy modes and use energy-efficient technologies like LoRa and NB-IoT (Sinha & Dhanalakshmi, 2022)	Higher energy consumption, especially in large-scale operations, makes them less eco-friendly (Boursianis et al., 2022)

(Continued)

TABLE 2.1 (*Continued*) Comparing IoT-based soil humidity measurement systems with conventional systems

CATEGORY	*IOT-BASED SYSTEMS*	*CONVENTIONAL SYSTEMS*
Environmental impact	Environmentally friendly. Many IoT sensors are solar-powered or designed for minimal energy use, promoting sustainability (Boursianis et al., 2022)	Traditional systems consume more energy and produce more waste, less ideal for eco-friendly applications (Sinha & Dhanalakshmi, 2022)
Operational cost	Lower long-term operational costs due to energy efficiency and automation, with minimal human intervention needed (Khaire & Dhanalakshmi, 2022)	Higher operational costs due to the need for manual data collection, calibration and regular equipment maintenance (Boursianis et al., 2022)
Automation and scalability	Highly scalable, supports automation, and can handle large-scale monitoring and management with minimal human oversight (Sinha & Dhanalakshmi, 2022)	Limited scalability due to manual processes and higher resource needs for large-scale implementation (Boursianis et al., 2022)
Suitability for large areas	Suitable for long-term and large-area monitoring due to automated, frequent data collection and real-time insights (Dutta et al., 2024)	It is less suited for large areas as manual interventions and limited data points reduce efficiency (Boursianis et al. 2022)

2.2.6 Accuracy and Precision: IoT vs. Conventional Systems

Conventional systems such as gravimetric analysis and tensiometers are well-recognized techniques for soil humidity measurement, providing accurate results in controlled settings. However, they are not ideal for long-term monitoring over large areas and are often prone to human interference, requiring frequent calibration (Boursianis et al. 2022). IoT-based systems, utilizing capacitive, resistive, and TDR sensors, offer a higher level of accuracy and precision in real-time. These systems continuously track variations in soil moisture, providing more frequent and refined data than what can be achieved manually (Khaire & Dhanalakshmi, 2022).

2.3 INTERNET OF THINGS (IoT) IN SOIL MONITORING

IoT has extended its applicability in many fields of engineering, including geotechnical engineering, where soil monitoring is considered to be most important by utilizing IoT technology. Over the literature mapping, it is clear that technological progress in IoT has enhanced the new way of monitoring our environment compared to the one used in the last decade. With respect to soil moisture, traditional techniques of estimation frequently fail to suffice, especially in the aspects of real-time measurement or frequent, large-scale surveys. However, IoT-based systems provide sophisticated, genuine, and real-time solutions that are competitive, innovative, flexible, and highly scalable in comparison with the existing ones in assessing the aspects of soil humidity content.

2.3.1 Components of an IoT-Based Soil Monitoring System

2.3.1.1 Sensors

Next, it is important to transfer this data to another system where it will be stored and analysed. IoT-based systems employ various network protocols, such as LoRa, Zigbee, and Narrowband IoT (NB-IoT), which enable stable, long-distance communications even in remote areas (Rehman et al., 2024). These protocols are energy-efficient, allowing sensors to operate for extended periods in the field.

Another crucial feature of smart IoT-based soil monitoring systems is their incorporation of cloud computing for soil data processing (Morchid et al., 2024). Data from the sensors are transferred, stored, and processed in the cloud, making it possible to store data cheaply, access it remotely, and incorporate sentiment analysis in real-time data processing. Cloud systems also allow engineers and farmers to model and predict future soil conditions based on past trends, streamlining the decision-making process. The complete structure of IoT-based soil monitoring system is shown in Figure 2.1.

2.3.1.2 Data Transmission

The integration of IoT sensors, data transmission, and cloud environments enables accurate soil humidity monitoring. The benefits of using IoT-based approaches over traditional methods include increased accuracy, speed, and scalability. Multiple sensors can be installed across large areas, with all data received into a single system, allowing for extensive environmental monitoring (Rajak et al., 2023).

IoT systems also provide accessibility through the World Wide Web, allowing stakeholders to make decisions from any location. For example, a construction manager on-site or a farmer overseeing a field of crops can manage soil conditions without constant monitoring, thanks to IoT systems (Rehman et al., 2024).

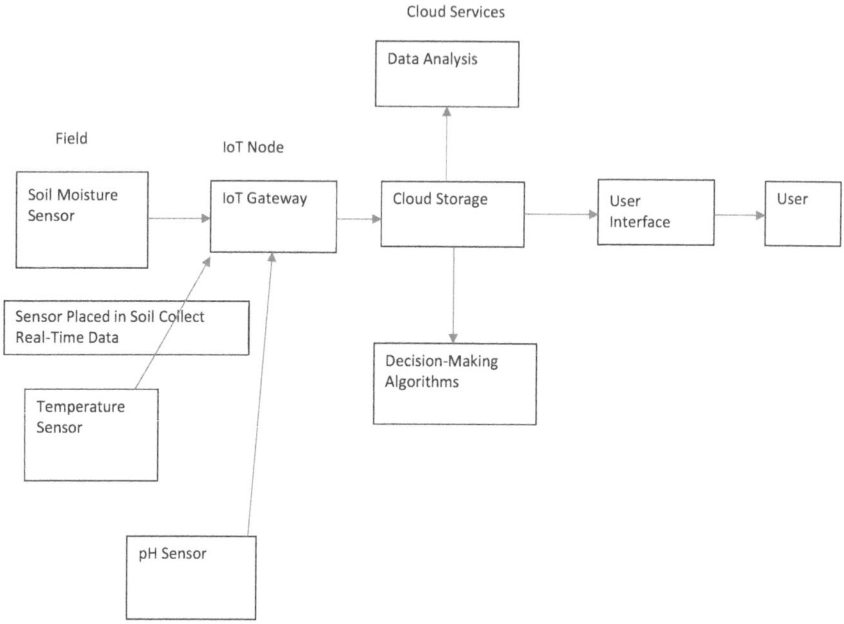

FIGURE 2.1 Structure of IoT-based soil monitoring system.

2.3.1.3 Cloud-based Data Processing

This also enhances real-time data analysis using cloud-based analytics, improving soil moisture control. For example, in agriculture, moisture levels can trigger dust control measures; or in construction, engineers can be alerted to potential land stability issues before disasters occur (Morchid et al., 2024).

Although IoT systems offer many benefits, there are challenges in adopting IoT systems. The initial costs of large-scale IoT implementation can be high, but long-term gains in efficiency, accuracy, and reduced labour costs often offset these expenses, making IoT the future solution for soil moisture monitoring (Tuan et al., 2020).

2.3.2 IoT-Based Soil Humidity Measurement Systems

Smart technologies which use IoT as a platform are widely used to measure soil humidity and are capable of supplying continuous real-time data, with which traditional methods are often unable to cope. One of the key components of these systems is that depending on various conditions and requirements of the plants, there are numerous types of sensors that are used for reading soil humidity.

Capacitive sensors are among the most frequent elements applied in the construction of IoT systems. Soil moisture has been sensed using these sensors through its detectable shift in the dielectric constant in the soil substrate. Capacitive sensors are usually relatively inexpensive, low power-consuming, and can remain accurate for a long time, which is why they can be used for agricultural as well as geotechnical purposes these days. Nevertheless, they can be sensitive to the type of soil in a particular area and may from time to time need to be calibrated.

$$V(t) = V_0\left[1 - e^{-\frac{t}{R_s C_s}}\right] + \frac{Z_{\text{TDR}}}{Z_{\text{TDR}} + Z_s} V_0 e^{-\frac{2t}{Z_{\text{TDR}} C_s}} \cos(\omega t - \phi)$$

$$\Delta Z_s = Z_s - Z_0 = \frac{\rho L}{A} - Z_0$$

$$C_s = \frac{\varepsilon A}{d}$$

$$V_{\text{reflected}}(t) = V_{\text{incident}}(t) \cdot \left(\frac{Z_s - Z_{\text{TDR}}}{Z_s + Z_{\text{TDR}}}\right)$$

$$T_{\text{delay}} = \frac{2L}{v}$$

$$v = \frac{1}{\sqrt{\varepsilon \mu}}$$

where

- $V(t)$ is the voltage as a function of time
- V_0 is the initial voltage applied to the sensor
- R_s is the resistance of the sensor
- C_s is the capacitance of the sensor
- Z_{TDR} is the characteristic impedance of the TDR system
- Z_s is the impedance of the sensor
- Z_0 is the reference impedance
- ρ is the resistivity of the material
- L is the length of the sensor
- A is the cross-sectional area of the sensor
- ε is the permittivity of the medium
- d is the distance between plates for the capacitive sensor
- ΔZ_s is the change in impedance of the sensor
- ω is the angular frequency
- ϕ is the phase shift
- T_{delay} is the time delay of the reflected signal
- v is the propagation speed of the signal in the medium
- μ is the permeability of the medium
- $V_{\text{reflected}}(t)$ is the reflected voltage as a function of time
- $V_{\text{incident}}(t)$ is the incident voltage as a function of time

2.3.3 Architecture of IoT Systems for Soil Humidity Assessment

2.3.3.1 *Sensor Networks*

In contrast, resistive sensors determine the electrical resistance of the soil – which tends to decline with increased moisture. These sensors are relatively simple and inexpensive to use but tend to degrade physically over time due to corrosion of the sensing material by aggressively acidic or alkaline solutions, including soils. This is, however, sometimes a limitation of resistive sensors as they commonly interfere with low-precision applications and yet are very affordable and powerful.

2.3.3.2 *Communication Protocols*

TDR sensors are a little more sophisticated and can provide very accurate measurements of soil moisture based on the amount of time it takes for an electromagnetic pulse to pass through the actual medium. Similar to what has been discussed with capacitive and resistive sensors, TDR sensors present low dependence on the soil type and are therefore well adapted to areas where heterogeneity of the soil is a problem. While TDR sensors are relatively costly, their higher accuracy of measurement cannot be overlooked, especially in very critical applications such as geotechnical or agricultural, where moisture content information plays a crucial role in decision-making.

2.3.4 Future Trends in Soil Humidity Monitoring with IoT

Following the advances in IoT, the application of smart sensors for monitoring soil humidity has remained dynamic, and promising IoT technologies have continued to enhance the sensors' accuracy, efficiency, and security. The development of IoT-based soil monitoring systems towards the future is in the way of integrating the latest smart sensors, artificial intelligence, machine learning, and blockchain technology to transform intelligent farming and construction sectors.

2.4 DATA COLLECTION AND ANALYSIS IN IoT SYSTEMS

The current construction of IoT systems for soil humidity assessment involves one or multiple sensor networks that collect data from different areas. Such sensors are established to cover large regions, resulting in highly detailed localized data on soil moisture.

The information gathered by these sensors is sent through different wireless interfaces, such as LoRa, Zigbee, and NB-IoT. Each of them has its advantages: the LoRa technology provides long-range communication, while NB-IoT consumes low power and is suitable for remote areas with few power sources. Of the two, Zigbee consumes more power but is ideal for transmitting huge amounts of data in the shortest time possible in areas with a high density of sensors.

Sensors and information gathering are important in IoT, and so is data processing to convert the collected data into useful information. Much of the collected data is transmitted to the cloud platforms where such data is analysed and processed. It is common that these platforms enable real-time monitoring of soil moisture and have extra features for pattern, trend, and anomaly detection applied to monitored data. Accompanied by this, the IoT system collects other parameters about temperature and humidity and offers a complete panorama of the condition of the soil, thus giving the user decision-making on irrigation, planning of land usage, or construction.

In agriculture, such a concept of collecting and utilizing big data facilitates better irrigation and usage of water and light for farming. It will ensure that soil stability is closely checked and that such disasters as landslides or failure in foundations are detected and corrected. Due to the highly developed sensors, reliable telecommunication connections, and extensive data processing methodologies in IoT-based Structural Health Monitoring System (SHMS), the current systems define new benchmarks highly surpassing the prior techniques for soil humidity monitoring.

2.4.1 Response Time and Real-Time Monitoring

The data acquisition and control capabilities of IoT-structured systems are significantly more advanced compared to conventional systems, which typically rely on manual measurements at fixed intervals. IoT systems provide continuous feedback on soil conditions, enabling timely responses to changes in moisture content. This is particularly useful in agriculture, where it helps optimize irrigation and prevent crop loss due to overwatering (Dutta et al., 2024). In geotechnical engineering, real-time monitoring helps identify changes in soil density early, preventing disasters like landslides or structural failures (Giannakidou et al., 2024).

2.4.2 Data Availability and Remote Access

Conventional systems require on-site field visits for data collection, which is time-consuming and often limited to specific sampling points. The data provided is typically a snapshot in time, offering limited insights into the dynamic behaviour of soil. In contrast, IoT systems provide real-time data acquisition, accessible remotely via cloud technologies (Feng, 2011). This capability allows users to manage and monitor soil humidity from any location, significantly enhancing data availability, especially in large-scale agricultural or construction settings (Boursianis et al., 2022).

2.5 IMPLEMENTATION CASE STUDIES

CASE STUDY 1: IoT-BASED SOIL MOISTURE MONITORING IN AGRICULTURE

For a large-scale smart farming project for efficient water management in an agricultural field, IoT-based soil moisture sensor system was installed on several hundred acres of land. To operate the system, capacitive sensors were installed in different sections of the fields, and a LoRa communication network ensured real-time transmission of obtained data. The information gathered by the sensors was collected and driven via an online interface and processed with weather patterns and cropping calendars. Precision horticulture information was immediately delivered to farmers for changes in the irrigation program as a result of the corrective measures on the moisture content of the soil. Thus, water consumption decreased by 30%, the effectiveness of watering increased, and the final crop yield improved as the energy usage was properly managed (Sinha & Dhanalakshmi, 2022). The IoT system helped enhance agricultural efficiency and sustainable farming, which was another added advantage in that it helped conserve water and energy.

CASE STUDY 2: APPLY IoT TO MEASURE THE HUMIDITY OF SOIL TO SUPPORT GEOTECHNICAL ENGINEERING VENTURES

In a real geotechnical engineering project of a construction of a significant large structure, IoT-based soil moisture sensors were used to measure the state of the ground. The site was situated in a potential landslide area and required varying stage monitoring to guarantee against this occurrence. TDR sensors in the network were used to record the amount of moisture within the soil at several layers. These sensors were tied to an online platform, which consisted of NB-IoT for long-range low-power data transfer. The system allowed constant checks on the findings and raised alarms if the corresponding moisture levels got to worrying ranges that pointed to stability loss. Due to the continuous flow of data, engineers were able to evaluate the probability of soil movement and then act accordingly in order to prevent construction loss or delay. Thus, the case under discussion demonstrates that IoT technology can promote safety and optimize geotechnical projects' efficiency while providing constant risk assessment.

CASE STUDY 3: COMPARATIVE STUDY OF IoT AND CONVENTIONAL SYSTEMS IN SOIL HUMIDITY MONITORING

In an experimental study to compare an IoT-based soil humidity system with a conventional system, two parcels of land were noted over a period of one semester. One plot used tensiometers and gypsum blocks to monitor the soil moisture levels, and the other one had IoT capacitive sensors. The traditional system involves the use of data that was gathered on a bi-weekly basis while the IoT sensors, on the other hand, relay data to a cloud platform in real-time. In the course of the study, the IoT system was able to give a more explicit view of the status of soil moisture and respond to the changes more proactively regarding irrigation. The earlier system, however, although precise, provided data less frequently to support timely changes and, therefore, over-irrigated during dry times and under-irrigated during wet times. The study revealed that IoT-based systems are more accurate, flexible, and efficient than analytical big data predictive-based systems and are more appropriate in massive application areas where real-time data is essential. This comparative analysis revealed the great potential of IoT to enhance the management of SoM, both in agricultural and geotechnical applications, provided that more dynamic and responsive systems than the traditional ones could be achieved.

2.6 CHALLENGES AND LIMITATIONS

Although a wide range of IoT-based systems has been presented for the monitoring of soil humidity, there are key issues and constraints that need to be overcome for its successful application. Such challenges posed technical and financial barriers, as discussed, that stall IoT technologies from being implemented broadly and effectively in different industries.

2.6.1 Challenges of Technical and Financial Nature in IoT Applications

One of the biggest challenges of IoT solution deployment is the cost of the hardware required to facilitate the deployment of the IoT, exclusive of sensors, data transfer devices, and the cloud. These initial costs can be expensive for small-scale farms or projects in particular, as seen above. IoT systems involve the use of technical interfaces for automation and need technical know-how in installation, configuration, and maintenance. Due to the emerging scales of interconnectivity, organizations require qualified personnel to deal with sensor networks and cloud computing services, and all of these hike costs. Moreover, some areas have issues with the IoT physical environment, that is, infrastructure that enables the efficient employment of the latest IoT technologies in less developed regions (Boursianis et al., 2022).

Thus, it can be said that the efficiency of such a system of IoT-based soil humidity control fully depends on the sensors. Some of the issues include the calibration of the sensors, which is considered an important factor because when the sensors are not well calibrated, they result in wrong data feedback that leads to wrong decisions. Soil differences, temperature, and any kind of environmental factors also hinder the correct working of humidity sensors and therefore cause inconsistency. This is because IoT systems must be monitored constantly, and their calibration must be updated on occasion to ensure that they deliver the best or desirable results, and this contributes to operational challenges. Another challenge is to maintain the stability of the sensors, especially their durability and effectiveness, in conditions that are unfriendly such as environments with high temperatures or high humidity.

2.6.2 Data Security and Privacy Concerns in IoT Systems

The readings also reveal that IoT systems use continually streaming data, which presents a challenge to the security and privacy of such data. This is so because the IoT sensors that gather data and stream it over the network may be compromised by hackers. Access by unauthorized people to this data poses immense dangers to the identity of the users; it also poses a high risk to the authenticity of the information used to make decisions. To counter the IoT systems' vulnerability, encryption practices, safe communication channels, and updating security regularly must be practised. However, the need for cybersecurity typically makes the system more complex and expensive, especially when monitoring is popular, as in remote or large-scale systems.

2.6.3 Energy and Network Limitations in Rural Regions

Most IoT systems operate in areas that have restricted access to electricity and consistent networks such as the internet. One of the issues that sensors located in some remote areas could have is power issues, as they may draw power from batteries or solar cells which are sometimes less efficient. While Low Power Wide Area Network (LPWAN) solutions such as LoRa and NB-IoT have advanced in providing remote sensor connectivity, there are issues with maintaining connections. Power limitations may adversely affect the durability and reliability of sensors, rendering more frequent maintenance or replacement necessary, and connectivity issues may cause data gaps and any real-time monitoring system to suffer in its functionality.

2.7 ADVANCES IN SENSOR TECHNOLOGIES FOR ENHANCED ACCURACY

Increased development in sensor technology means that the scope, accuracy, and durability of soil humidity monitoring will improve in the subsequent years. Future sensors will therefore require higher miniaturization, reduced power consumption, and will be more sensitive to different environments. The development of a multi-parameter sensor that not only tracks soil humidity but also other factors such as temperature, salinity, and pH will advance real-time information gathering to the next level. These sensors will provide a better assessment of the state of soils, which is particularly advantageous for applications across the agricultural and geotechnical industries.

2.8 INTEGRATION OF AI AND MACHINE LEARNING FOR PREDICTIVE SOIL ANALYSIS

Artificial intelligence (AI) and machine learning (ML) have now become significant drivers for the future of soil humidity monitoring. ML can then be used with gateways to generate alerts on the data processed from sensors so that real-time analytics can be applied to what could be predicted. For instance, by analysing the existing data on soil moisture and weather conditions, growth cycle of crops, and existing and future data, the ML algorithms can make a precise prediction regarding the soil moisture levels that will help to make proper decisions on irrigation and management of the land territory. For example, future soil forecasts will lead to effective water utilization as well as proper management of soil with less harm to the environment as well as conserving the financial resources from the side of the farmers and engineers.

2.9 POTENTIAL USE OF BLOCKCHAIN FOR SECURE DATA MANAGEMENT IN IoT SYSTEMS

The integration of blockchain in IoT systems will make the management of huge data collected on ground sensing devices more secure at the same time. Blockchain will allow for all data, ranging from the information collected by the sensors to the action undertaken, to be recorded and immune to forgery. This dispersed manner of focusing on data management will thus enhance credibility and openness, especially where data integrity plays an important role, including in construction and agriculture. Blockchain can also assist smart contracts in IoT-based systems like soil monitoring systems to perform actions concerning predefined rules or conditions like changes in irrigation depending on the quantities of moisture in the soil.

2.10 CONCLUSION

In soil, humidity has its importance in geotechnical engineering as well as agricultural sciences, in which determination of the humidity level in the soil is capable of exerting a considerable impact on the results of the constructed facilities and cultivated plants. Some of the most typical historic approaches to determining soil moisture include gravimetric measuring, tensiometers, and gypsum blocks. As much as these techniques provide reliability, they have some inherent disadvantages of inaccuracy, real-time determination, and ease of use.

IoT technology brings an exciting prospect in the manner soil monitoring has been conducted in the past. IoT-based systems offer data collection and monitoring from a distance and greater accuracy than traditional methods. By integrating more precise sensor networks, reliable signal transmission techniques, and utilizing cloud computing in data storage and analysis, these systems are radically changing the ways that soil moisture evaluation is done. Capacitive sensors, resistive sensors, and TDR are the technologies in use to enable this transformation.

Comparing IoT systems with conventional techniques reveals numerous advantages: more precise results, immediate feedback, higher accessibility of the data, and higher energy efficiency. By analysing different cases, IoT has proved its applicability in the management of soil moisture in agricultural and geotechnical application fields.

However, there are still obstacles, including technical issues, financial constraints, sensor calibration, and data security issues. Nonetheless, future advances in sensor and drive technologies as well as the use of AI predictive analyses or even block-chained smart sensors are already on the horizon. While smart construction and smart farming are integrative of IoT, innovation is still continuous, and thus it is a promising area.

REFERENCES

Boursianis, A. D., Papadopoulou, M. S., Diamantoulakis, P., Liopa-Tsakalidi, A., Barouchas, P., Salahas, G., & Goudos, S. K. (2022). Internet of things (IoT) and agricultural unmanned aerial vehicles (UAVs) in smart farming: A comprehensive review. *Internet of Things, 18*, 100187.

Darsi, B. P., Molugaram, K., & Madiraju, S. V. H. (2021). Subgrade black cotton soil stabilization using ground granulated blast-furnace slag (GGBS) and lime, an inorganic mineral. *Environmental Sciences Proceedings, 6*(1), 15.

Dutta, S., Adhikary, S., Bhattacharya, S., Roy, D., Chatterjee, S., Chakraborty, A.,... Rajak, P. (2024). Contamination of textile dyes in aquatic environment: Adverse impacts on aquatic ecosystem and human health, and its management using bioremediation. *Journal of Environmental Management, 353*, 120103.

Feng, Z. (2011). Research on water-saving irrigation automatic control system based on internet of things. In *International Conference on Electric Information and Control Engineering* (pp. 2541–2544). Romania, Iaşi: IEEE.

Giannakidou, S., Radoglou-Grammatikis, P., Lagkas, T., Argyriou, V., Goudos, S., Markakis, E. K., & Sarigiannidis, P. (2024). Leveraging the power of internet of things and artificial intelligence in forest fire prevention, detection, and restoration: A comprehensive survey. *Internet of Things, 26*, 101171.

Gopikumar, S., Banu, J. R., Robinson, Y. H., Shanmuganathan, V., Kadry, S., & Rho, S. (2021). Novel framework of GIS based automated monitoring process on environmental biodegradability and risk analysis using Internet of Things. *Environmental Research, 194*, 110621.

Khaire, U. M., & Dhanalakshmi, R. (2022). Stability of feature selection algorithm: A review. *Journal of King Saud University-Computer and Information Sciences, 34*(4), 1060–1073.

Krishnan, R. S., Julie, E. G., Robinson, Y. H., Raja, S., Kumar, R., & Thong, P. H. (2020). Fuzzy logic based smart irrigation system using internet of things. *Journal of Cleaner Production, 252*, 119902.

Morchid, A., Alblushi, I. G. M., Khalid, H. M., El Alami, R., Sitaramanan, S. R., & Muyeen, S. M. (2024). High-technology agriculture system to enhance food security: A concept of smart irrigation system using Internet of Things and cloud computing. *Journal of the Saudi Society of Agricultural Sciences, 145*, 1–17.

Pamula, A. S., Ravilla, A., & Madiraju, S. V. H. (2022). Applications of the internet of things (IoT) in real-time monitoring of contaminants in the air, water, and soil. *Engineering Proceedings, 27*(1), 26.

Rajak, P., Ganguly, A., Adhikary, S., & Bhattacharya, S. (2023). Internet of Things and smart sensors in agriculture: Scopes and challenges. *Journal of Agriculture and Food Research, 14*, 100776.

Ravilla, A., Shirkey, G., Chen, J., Jarchow, M., Stary, O., & Celik, I. (2024). Techno-economic and life cycle assessment of agrivoltaic system (AVS) designs. *Science of the Total Environment, 912*, 169274.

Rehman, A., Alblushi, I. G. M., Zia, M. F., Khalid, H. M., Inayat, U., Benbouzid, M.,... Hussain, G. A. (2024). A solar-powered multi-functional portable charging device (SPMFPCD) with internet-of-things (IoT)-based real-time monitoring—An innovative scheme towards energy access and management. *Green Technologies and Sustainability, 3*, 100134.

Sinha, B. B., & Dhanalakshmi, R. (2022). Recent advancements and challenges of Internet of Things in smart agriculture: A survey. *Future Generation Computer Systems, 126*, 169–184.

Tuan, T. A., Long, H. V., Son, L. H., Kumar, R., Priyadarshini, I., & Son, N. T. K. (2020). Performance evaluation of Botnet DDoS attack detection using machine learning. *Evolutionary Intelligence, 13*(2), 283–294.

Soil Statistics Anthology for Investigation and Relevance Using IoT

3

3.1 INTRODUCTION TO SOIL STATISTICS AND IoT INTEGRATION

There is a groundbreaking synergy between soil statistics and Internet of Things (IoT) and both concepts are exciting additions to geotechnical engineering (Borrelli et al 2017). The nature and quality of soils are of paramount importance to any construction as soil data is used to determine the stability of the structures (Johnson-Bailey et al. 2009). In the past, work on the determination of soil conditions involved extensive us of hand augers and samplers. On the same note, with the soon-to-be near-universal us of IoT technology, data collection is now quicker, more efficient, and can be done i real-time in some cases (Anupam et al., 2015). This change has shifted how engineer use soils for analysis and design work, thus making it less complicated and more depen dent on research data (Najjar et al., 2024).

Statistics of soil are helpful in establishing some important characteristics of soil concerning construction activities. These include moisture content variation, textur variation, and density variation, all of which affect the response of structures con structed on the sub-soil (Dobos et al., 2001). By using computers, engineers work ou these variations, and the performance of soil under given conditions is estimated. Thus this statistical technique allows for finer planning of the construction of geotechnica structures, preventing assumptions regarding the behaviour of the ground from bein too daring and creating fundamentally more reliable structures (Helber et al., 2023).

DOI: 10.1201/9781003537625-

The place of IoT in current soil investigations is revolutionary. Many IoT devices are also placed on the ground, and they constantly record and deliver data on soil conditions to engineers. Some of it can be temperature, moisture content, and the movement of the soil, which helps in tracking stabilities over time (Jeffries et al., 2020). IoT technology does not require traditional, systematic data gathering and collection techniques, which are more mechanical in their approach to soil investigation. This information can be analysed immediately, thus issuing early signs of possible risks (Burdett & Wellen, 2022).

It is noteworthy that the primary goal of IoT applications for the analysis of soil data is improving the accuracy of investigations and reducing the time for investigations on the soil (Helber et al., 2023). Soil analysis has, for many years, depended on conventional techniques unlike IoT, which continually monitors and quickly analyses data. This technological advancement is well applied where there are varying soil conditions that change with time in large projects (Jeffries et al., 2020). The data received through IoT systems not only enhances the knowledge of soil behaviour but also contributes to the creation of forecast models that will help make construction work safer and less costly (Najjar et al., 2024).

Statistical representation is useful for quantifying the properties of soils, as knowledge of these properties is fundamental to the assessment and execution of any geotechnical project (Borrelli et al., 2017). Essentially, soil descriptors, including particle size distribution, bulk density, void ratio, and water suction, are essential in evaluating their stability and constructability. Using statistical techniques, engineers can express these properties in forms amenable to comparison, recognition of trends, and prediction of soil response under different conditions (Anupam et al., 2015).

That is why the problems of soil classification and the determination of specific types of soil are considered important by the users of soil materials. Such classifications include sandy, clayey, loamy, and silty soils, each of which has characteristics that influence construction processes (Dobos et al., 2001). These classificatory divisions are based on particle size, grain mineral content, and organic composition, which in turn affect characteristics such as permeability, compressibility, and strength (Johnson-Bailey et al., 2009). These classifications describe the mean values and provide statistical information regarding variability within these categories to predict which soil type is more appropriate for use in any given geotechnical application (Borrelli et al., 2017).

Therefore, the parameters that define the soil characteristics of interest for geotechnical investigations include texture, density, and water content (Najjar et al., 2024). Soil is characterized by its texture, which determines its water-holding capacity and permeability. On the other hand, density determines the nature of compaction and the soil's bearing capacity (Jeffries et al., 2020). Soil consistency and moisture content are important as they impact shear strength and bearing capacity. By measuring these properties, engineers can determine the soil's validity in supporting structures and design foundations to reduce structural risks (Helber et al., 2023).

Probability concepts are used to assess quantitative aspects of soil data and to make quantifiable evaluations of soil behaviour (Burdett & Wellen, 2022). Measures such as the mean, median, and standard deviation help describe summary statistics of soil properties, while more complex techniques like regression and prediction models forecast soil behaviour under specific loads and conditions (Najjar et al., 2024). These statistical tools are critical in recognizing patterns, trends, and risks and establishing structures that factor in soil properties (Helber et al., 2023).

IoT sensors are central to contemporary soil investigation procedures. They can be embedded into the soil to record variables like temperature, moisture, and pressure, which provide frequent readings (Jeffries et al., 2020). These sensors, connected wirelessly to monitoring systems, allow real-time monitoring of soil conditions, enabling engineers to identify potential issues such as changes in moisture content that may cause instability or failure (Borrelli et al., 2017).

General approaches to using IoT sensors in soil investigations convey crucial inputs on data collection (Johnson-Bailey et al., 2009). It is important to place sensors at appropriate locations depending on the specific project requirements. For instance, sensors might be placed at different depths to provide a comprehensive picture of the soil state (Borrelli et al., 2017). When using IoT sensors, environmental conditions, such as weather and geographical terrain, which can affect both the soil and the sensors' operational capabilities, should also be considered (Anupam et al., 2015).

3.2 FUNDAMENTALS OF SOIL PROPERTIES AND THEIR STATISTICAL REPRESENTATION

The properties of soil constitute a fundamental area of geotechnical engineering, whereby one learns or expects the behaviour of soil in specific conditions. Efficient representation and analysis of these properties are critical in engineering safe and efficient infrastructures (Ma et al., 2024). This view is due to the complexity of soils and their characteristics, which differ between types and classifications.

Possible classifications of soils include those that result from their physical and chemical characteristics. According to texture, the four most familiar types include sand, clay, silt, and loam (Pathak et al., 2023). For instance, group "E" is characterized by having large particles, leading to high permeability of sand and low resistance to bearing structures. Sandy and clayey soils, on the other hand, are fine-grained soils that in particular contain water, which may also swell or shrink depending on the conditions, forming potential problems for the stability of foundations. The silt soils are intermediate between sand and clay in water retention and permeability, and loamy soils provide the right mix of sand, silt, and clay, which is ideal for construction (Ansarifar et al., 2021). These classifications are crucial in determining the characteristics of soil when in a loaded state and in determining the most appropriate types of soil for use in any given geotechnical application.

There are several "factors of influence" that cause changes to soil properties and mechanical characteristics suitable for construction (Ma et al., 2024). Relative proportions of sand, silt, and clay, collectively referred to as texture, control how a particular type of soil allows water to pass through and how it retains water. This property is referred to as bulk density and relates to the soil's strength and load-bearing ability. High-density soils offer relatively more stability, whereas low-density soils may be susceptible to compaction or reinforcement (Pathak et al., 2023). Another important factor is moisture content, which determines the soil's strength and volume change in the presence of water. Fluctuations in moisture content can cause expansion, contraction, or changes in bearing capacity, making moisture content monitoring critical during construction (Ansarifar et al., 2021).

Other important properties include porosity, which defines the degree of permeability, and compaction, which measures the added weight soil can comfortably hold. Understanding these properties enables geotechnical engineers to make informed decisions on site suitability and necessary soil treatments (Ma et al., 2024).

3.2.1 Statistical Techniques for Soil Property Analysis

Statistical analysis is critically important in interpreting soil properties and behaviours based on them. Common techniques include mean, median, and standard deviation, which help summarize average values for properties like soil moisture or density (Ansarifar et al., 2021). More advanced methods such as regression analysis and correlation studies can define relationships between soil properties, like moisture content's impact on compaction or stability (Pathak et al., 2023).

Machine learning (ML) and artificial intelligence (AI) tools have recently been incorporated into soil assessments. These technologies model complex cognitive relationships and handle large datasets from IoT sensors, offering real-time insights into soil properties (Ma et al., 2024). Probabilistic and AI-based methods provide an improved understanding of soil behaviour, making them essential in contemporary geotechnical engineering (Pathak et al., 2023).

3.2.2 Statistical Relevance and Interpretation in Geotechnical Engineering

Analysis of soil data is of central importance to geotechnical engineering for the effectiveness and security of civil engineering projects. Through this acute analysis, engineers may make decisions, evaluate data obtained from IoT devices, and gain useful knowledge from the measurements regarding the soil. This section looks at the need for statistical analysis, the effect of accurate data on structural safety, and the approach of verifying and validating data on the soil obtained via the IoT sensors. The usefulness of statistical analysis for soil information is applied in civil engineering projects. Data analysis with regards to soil, or more broadly statistical analysis, is an important factor necessary for planning and undertaking civil engineering projects. By applying robust statistical techniques, engineers may make informed decisions, assess the accuracy of data generated by IoT devices, and gain insightful knowledge from soil measurements.

This section examines the importance of statistical analysis, the impact of accurate data on structural safety, and the methods for validating and verifying soil data collected by IoT sensors.

3.2.2.1 *The value of statistical evaluation for soil data in civil engineering projects*

- **Understanding and Analysing Soil Data**: Basically, the area of statistical analysis is crucial for planning and executing engineering projects. Depending on the depth and location, the soil's properties can vary significantly, including its shear strength, density, and moisture content. By applying statistical techniques, engineers can more effectively summarize these variances and identify patterns that inform design decisions.
- Descriptive statistics, which comprise measurements of central tendency (mean, median, mode) and dispersion (variance, standard deviation), give a complete picture of soil qualities. These measurements are also required in order to calculate representative mean soil values and to understand the nature of soil variation. Use of inferential statistic means that engineers can be able to build slopes, foundations and any other structures with accuracy since they are in a position to generalize and estimate on the behaviour of the soil from the sample findings.
- By applying robust statistical techniques, engineers may make informed decisions, assess the accuracy of data generated by IoT devices, and gain insightful knowledge from soil measurements. This section examines the importance of statistical analysis, the impact of accurate data on structural safety, and the methods for validating and verifying soil data collected by IoT sensors.
- Statistical methods can help in assessing uncertainty and risk. For example, probabilistic models can be employed to estimate the risk that built-up soils or grounds will give way or deform in different loading conditions to facilitate the construction of safer and more durable structures. Statistics in effect translates raw soil information into useful data that makes engineering decisions independent of physical soil characteristics possible.
- **The Connection between Accurate Data and Structural Performance and Safety**: Civil engineering projects and their safety depend significantly on the accuracy of soil data, which directly influences the stability, durability, and overall performance of structures, ensuring compliance with engineering standards and minimizing potential risks. It is only possible to collect true structures through precise soil data; otherwise, structures founded on the ground will only endure the impacts and loads that it will be subjected to. Failure to accurately estimate the load-bearing capacity due to wrong or inadequate information on the soil might cause building collapse, very high costs of maintenance, or the creation of an unsafe environment.

- For instance, there is a need to consider some properties of soils like bearing capacity and settlement characteristics while designing the foundation because this takes into consideration the capacity of the foundations to bear loads without unduly settling or collapsing and the stresses they will encounter. Underestimating load-bearing capability as a result of faulty or insufficient soil data could lead to structural failure, excessive maintenance costs, or hazardous conditions.

  For example, carefully considering soil properties, such as bearing capacity and settlement characteristics, is crucial to foundation design since it ensures that the foundations can support anticipated loads without excessively settling or collapsing. When constructing measures for slope stability, accurate information in relation to the ground also ensures that not only cuttings and embankments are stable in a variety of conditions but also affect the operation of the structure. It affects the durability of structures; it shapes construction processes in relation to time and financial perspectives on engineering concepts. Hence, the calls for the appropriateness of soil data by civil engineers to facilitate and attain favourable project performance and credibility of civil engineering projects.
- **Evaluation and Verification of Data from IoT-Collected Soils**: As the approach to track soil information, IoT has been widely used; consequently, certain data collected by such sensors has to be checked and approved. Nonetheless, IoT sensors provide promising advancements, but they might have some problems like sensor failures, data abnormalities, and incorrect calibration.
- **Upkeep and Calibration**: The data being generated by IoT sensors is correct; hence, it is essential to adjust these sensors periodically. Some of the calibration processes involve comparing the data obtained from the sensors with set standards and the alteration of the data if needed. As important as they are for routine inspections and maintenance, checking for and then repairing any probable problems which may pose a threat to the data and information accuracy are also valuable.
- **Systems for Intelligent Soil Management**: Possible future advancements may result in the emergence of intelligent soil management systems linking IoT with various decision-support tools. Such systems could be designed to provide real-time decision support on issues related to management, monitoring, and intervening with respect to the soil data collected.
- **Networks for Adaptive Monitoring**: In time, IoT devices will be capable of altering how they monitor incoming data and interactions they are exposed to. These networks will facilitate effective and timely soil monitoring and enhance the quality of soil-based studies.

3.2.3 Leveraging IoT for the Development of New Tools for Soil Analysis

As IoT continues to evolve, it is expected that methodologies of studying the soil wi greatly transform.

- **Sensor Technology of the Future**: As the development of the sensors increases, more advanced, accurate, and much cheaper sensors are manufactured. Future sensors will likely be even more precise and sensitive with greater operational distance and insensitivity to modifications around them. Nanosensors like the RFID Smart Dust and wireless sensor networks (WSNs) are examples of newer ways through which data collection and monitoring would be feasible.
- **Self-Governing Monitoring Systems**: IoT with robotic sensors and drones shall enable high resolution of several angles of the soil. While drones with sophisticated sensors can fly and gather data on the environment, often over large areas, robotic systems are able to perform measurements and inspections at the scene.
- **Energy-Efficient IoT Devices**: The call for more or continuous soil monitoring means that the establishment of efficient Internet of Things sensors for energy will be more and more important. Due to low-power sensors and energy harvesting schemes, data acquisition would be long-term and require minimal intervention in terms of maintenance and battery change. Solutions in the progress of the soil data machine learning and analytics ML is a growing and dynamic field which has new techniques to analyse data and forecast the behaviour of the soil.
- **Advanced Statistical Models**: Recent advances will improve the mimicking of complex interactions and soil dynamics. That is why, employing methods such as multivariate analysis or spatial statistics will help to identify the nature of soils and the changes in their nature. With improved statistical methods, the accuracy of forecasting pertaining to other issues, such as soil stability, settlement, and so on, will also increase.
- **Algorithms for Machine Learning**: In other words, massive-scale soil data analysis with options from ML is increasingly feasible. Supervised learning, unsupervised learning, and deep learning enable the build-up of more accurate predictive models of the behaviour of soils. The discovery of these algorithms is possible about the connections and relations that traditional statistical techniques may not see.
- **Big Data Analytics**: The real-time data gathering made possible by IoT integrated with big data analysis will address the inherent issues of data flood that come with soil monitoring activities. This will enhance real-time analysis and response to data from the domain along with a precise decision concerning the use of soil.

3.2.3.1 New directions for Internet of Things for soil sensing applications

- A range of potential future enhancements in IoT applications for soil monitoring could entirely transform the ways of gathering, assessing, and using soil data.
- **Integration of Augmented Reality (AR)**: This is made possible through AR technology, which provides engineers with a live view of soils so they can get a glimpse of how it feels to work on them. If done this way, it could lead to a better description of the soil properties and raise the bar on what constitutes a geotechnical investigation.
- In the dissemination of the blockchain, the soil data collected by IoT was to be protected and safeguarded for confidentiality and integrity. Thus, applying blockchain technology in the ongoing scientific analysis of soil samples can help improve the trustworthiness of the data produced by translating those differences into a non-interpretable blockchain system.
- **Access Control**: Tight access control protocols ensure that the data can only be viewed and handled by authorized users. Role-based access control (RBAC) and multi-factor authentication (MFA) may strengthen security and reduce the likelihood of unauthorized access.
- **Regular Audits**: Security audits and vulnerability assessments are regular ways to find potential weaknesses in IoT systems. Addressing these vulnerabilities as soon as possible could help prevent security lapses and ensure the integrity of the data going forward.
- The results of traditional soil sampling and laboratory testing can be used to cross-validate IoT data. IoT data provenance verifies the truthfulness of the data, and any inconsistency that may call for intervention is identified during cross-validation. One of the ways of validating an IoT sensor is by comparing the data retrieved by the sensor with that obtained from soil tests conducted in the laboratory.
- When implementations of quality control measures are in place, then one is able to monitor variation and inconsistencies of soil governance structures that integrate data from IoT with decision-support tools. These systems may offer automated recommendations for soil management, monitoring, and intervention based on real-time data analysis.
- **Networks for Adaptive Monitoring**: As IoT devices become more widespread, so will their ability to adapt, allowing them to change their monitoring strategies in response to real-time data and outside influences. These networks will enable dynamic and responsive soil monitoring, optimizing data collection and improving the accuracy of soil assessments.
- **Access Control**: Tight access control protocols ensure that the data can only be viewed and handled by those who are authorized. RBAC and MFA may strengthen security and reduce the likelihood of unauthorized access.
- **Regular Audits**: Security audits and vulnerability assessments are regular ways to find potential weaknesses in IoT systems. Addressing these

vulnerabilities as soon as possible could help prevent security lapses and ensure the integrity of the data going forward.

- The results of traditional soil sampling and laboratory testing can be used to cross-validate IoT data. The veracity of IoT data is confirmed, and any discrepancies that could require attention are discovered through cross-validation. Comparing an IoT sensor's real-time soil moisture data with lab tests is a useful method of confirming the sensor's accuracy.
- When quality control techniques are applied, anomalies and discrepancies in the data are monitored. A few more cases include too low/ too high values, which may be a result of external interference or incorrect sensor measurements. Measures of data cleansing and sampling are used to handle or eliminate such irregularities.

3.3 IoT SENSOR TECHNOLOGY FOR SOIL INVESTIGATION

IoT sensor is a relatively new technology in the soil investigation process, but it has enhanced data collection on soil conditions by geotechnical engineers and provided real-time data for observation. Sensors are combined with the wireless communication system for effective, accurate, and less invasive gathering of important properties of soils. This technology is used to gain information to understand the behaviour of the soil that can be used to prevent or anticipate geotechnical failures. Therefore, using IoT in soil investigations enables engineers to make decisions with higher accuracy and within the shortest time.

3.3.1 Types of IoT Sensors for Soil Monitoring

There are various IoT sensors for use in monitoring the soil, and each kind of IoT sensor has the capability of computing characteristics that define the performance of soil. Some of the most commonly used sensors include the following.

- **Soil Moisture Sensors**: These sensors measure the volume of water in the soil, which is always a significant feature when studying the impact of water on soil strength and stability. They are installed in areas that experience floods or droughts because the soils usually fluctuate a lot with water levels.
- **Temperature Sensors**: This chapter demonstrated that soil temperature affected the physical and chemical characteristics of the soil. Thermometers utilized in monitoring the changes in temperatures allow for the tracking of fluctuations in the rate at which the soil can hold water, the rate at which microbes deteriorate, and the compressive force of the ground.

- **Pressure Sensors**: These sensors assess the pressure and stress imposed on the soil for evaluating the response of soil in a loaded state, such as foundations, retaining walls, embankments, and so on. This information is relevant in any settlement and any other structural transformation processes.
- **Electrochemical Sensors**: These are used to quantify changes in no response and to determine the chemistry of the soil for its ability to produce fertility, corrosion, and other environmental factors. They are very much used in areas such as agri-business/commodity production and physical science, especially ecology.
- **Strain Gauges and Tilt Sensors**: These sensors monitor ground displacements and deformation of the soil, which are very essential in predicting potential slide areas, stability of slopes, and differential movements or settlings as they happen. Despite getting the data by applying different types of sensors, geotechnical engineers make valuable conclusions about the dynamic and static state of the soil in order to provide an analysis of possible risks and actions.

3.3.2 Deployment Strategies for Soil Investigation

This means that the deployment of IoT sensors in soil investigation is only as good as the planning done in it. However, for sensor installation, a site survey is necessary to choose appropriate areas for engineers to receive data. From these observations, therefore, the type and quality of the soil, state of the environment, and general conditions of the project define the overall method and approach to deploying the sensors.

In general, at least two to three sensors need to be inserted into the ground as a rule to gain an insight into the condition of the soil. For instance, there are sensors in contact with the soil, as shallow sensors can measure the moisture content or the temperature on the surface of the soil while deep sensors can measure conditions such as pressure and soil density within the subsurface. However, the use of this layered approach is overall more comprehensive, and I believe that this adds another needed layer of insight into the behaviour of the soil.

Other deployment strategies also have to do with aspects of the project site structures. For example, in slope stability monitoring, it is demonstrated that sensors are placed along the slip line to the earliest sign of any instability. For instance, it is possible to install several sensors at several intervals to monitor territorial areas in the cities to measure ground movements or subsurface movement.

Wireless communication, as against wired communication, is built into the provision of IoT sensors because data has to be relayed from places that are fully off-grid or hard to reach. LoRaWAN, Zigbee, and cellular interfaces are among the wireless communication interfaces used most, based on data transmission and environmental demands. Such communication networks enhance the transfer of data each time the facility in which the patient is located is entered and reduce frequent on-site visits for constant monitoring.

3.3.3 Real-Time Data Acquisition and Monitoring

Real-time data is perhaps one of the most significant benefits that IoT provides in the field of soil investigation. Fixed monitoring approaches, such as periodic sampling and soil testing, yield limited information on soil conditions since it overlooks most of the changes in the conditions. As for IoT sensors, these are useful for monitoring the properties of the soil in real-time, as they are unhelpful when it comes to the location of objects on the construction site.

Real-time data acquisition is useful, especially when the status of the ground changes quickly, for instance, during the rainy season, earthquake, or construction phase. Engineers can get notified as soon as the moisture content of the soil is high and beyond the safe level, pressure is higher than the allowed limit, or there is movement on the soil. It also reveals real-time information that can assist engineers in responding to the situation before a part fails, which may be cost- and time-saving.

Furthermore, the data gathered by IoT sensors can be processed and reflected on inbuilt software platforms. Such platforms enable engineers to analyse trends, do predictive analysis, and thus predict further behaviour of soil. Significant data concerning events like soil liquefaction or slope failure can also be entered, and automated systems can then generate alerts or schemes for intervention when sympathetic signals are detected.

3.3.4 Soil Sampling and Data Collection Techniques Using IoT

Soil sampling is one of the fundamental activities in geotechnical engineering and determines the required data on soil characteristics. In the past, soil sampling has been done using a process that involved physically taking samples from the soil, transporting them to laboratories, and then analysing them for a number of values, including moisture, density, and so on. However, with the developments in technology, specifically in IoT, soil sampling and data collection have not been the same again, considering speed, accuracy, and real-time data collected. This transformation improves the quality and availability of the information, which plays a crucial role in the decision-making process in construction, agriculture, and environment monitoring.

3.3.5 Traditional versus IoT-Enhanced Soil Sampling Methods

The direct collection of samples from the ground is done using augers or shovels at several depths and points. These samples are then taken to a laboratory to test physical properties that include moisture content, pH, texture, and content. Although helpful, it is highly manual, lengthy, and prone to deferred responses in a subsequent analysis of esults. Furthermore, random sampling of soil requires a periodic timeframe, which esults in some changes in soil conditions that may not be realized until the next sam-ling period.

On the other hand, IoT-enabled soil sampling reforms this task through the incor-oration of smart sensors that enable constant and real-time collection of soil properties. n other words, it is possible to monitor soil conditions constantly without necessar-ly having to take samples quite often due to IoT. Real-time monitoring offers more ive information and enables engineers and scientists to respond to the changes in soil ehaviour like variations in moisture content, pressure, and likely instability.

Additionally, systems based on IoT can gather data from different places and posi-ions to analyse soil conditions in several points of interest areas. This approach also ets rid of so many logistic issues which are usually encountered with traditional meth-ds and has a good impact of increasing the reliability of the collected data. Traditional oil sampling methods versus IoT-enhanced soil sampling are shown in Table 3.1.

TABLE 3.1 Traditional soil sampling methods versus IoT-enhanced soil sampling

DIMENSION	*TRADITIONAL SOIL SAMPLING METHODS*	*IOT-ENHANCED SOIL SAMPLING METHODS*
Cost	• Relatively high labour costs due to manual sampling and transportation of samples to labs	• Initial setup costs can be high due to sensor deployment, but operational costs are significantly lower with automated data collection and fewer human interventions
Accuracy	• Moderate accuracy, often limited by the number of samples and the manual interpretation of results	• High accuracy with real-time, continuous monitoring of multiple soil parameters (e.g., pH, moisture, temperature) through deployed sensors across different locations
Time efficiency	• Time-consuming; sampling requires manual labour, and laboratory analysis may take days or weeks	• Significantly faster as sensors provide real-time data that can be immediately analysed without waiting for lab results
Data collection frequency	• Limited to scheduled intervals when human operators collect samples	• Continuous data collection with IoT sensors, providing insights into soil health and conditions on a near-constant basis
Scalability	• Not easily scalable; requires more resources (human labour, time) to cover larger areas	• Highly scalable; once sensors are deployed, large areas can be monitored with minimal additional costs or effort

(*Continued*)

TABLE 3.1 (*Continued*) Traditional soil sampling methods versus IoT-enhanced soil sampling

DIMENSION	*TRADITIONAL SOIL SAMPLING METHODS*	*IOT-ENHANCED SOIL SAMPLING METHODS*
Geographical coverage	• Limited coverage; sampling typically focusses on specific points, which may not represent the entire area accurately	• Broad geographical coverage with multiple sensors across various locations, offering a comprehensiv view of soil conditions across large fields or diverse terrains
Ease of use	• Requires trained personnel to collect and analyse samples, leading to a steep learning curve	• User-friendly; many IoT systems come with interfaces and dashboards for easy monitoring an interpretation, with minimal trainir needed for basic operation
Data analysis	• Manual data entry and analysis, prone to human errors, with limited analytics capabilities	• Automated data logging and advanced analytics tools like machine learning models can dete patterns, provide forecasts, and improve decision-making on soil management
Real-time monitoring	• Not possible; samples are processed in batches and results take time to arrive	• Real-time monitoring allows for instantaneous insights into soil conditions and immediate respons to changes, such as irrigation need
Environmental impact	• Increased environmental footprint due to manual processes, vehicle use for sample collection, and lab-based chemical testing	• Reduced environmental impact wit fewer emissions from manual processes and the ability to detect and address environmental issues like soil erosion or pollution in real-time
Adaptability to changing conditions	• Slow to adapt; changes in soil conditions may go unnoticed for days or weeks between sampling sessions	• Highly adaptive; sensors can detect and respond to changes in soil conditions immediately, enabling proactive measures such as adjustir irrigation or applying fertilizers bas on real-time data
Data storage	• Physical storage (e.g., paper, lab reports) and manual records, making long-term tracking difficult	• Digital storage with cloud-based platforms allows for easy access to historical data and trend analysis over time, facilitating better decision-making and planning

(*Continued*)

TABLE 3.1 (*Continued*) Traditional soil sampling methods versus IoT-enhanced soil sampling

DIMENSION	*TRADITIONAL SOIL SAMPLING METHODS*	*IOT-ENHANCED SOIL SAMPLING METHODS*
Sampling resolution	• Low resolution, typically based on fewer sampling points, which might not reflect variations in soil properties across the field	• High resolution with dense sensor networks, providing granular insights into local soil variations and offering more precise data to farmers or researchers
Maintenance requirements	• Little to no ongoing maintenance for the sampling process itself, but lab equipment may require upkeep	• Sensors require regular maintenance, including battery replacements or software updates, but can operate autonomously for long periods with minimal intervention
Human error	• High potential for human error during sample collection, transportation, and analysis	• Minimal human intervention, reducing the chance for human error in data collection and interpretation, though sensor calibration and placement accuracy are important
Labour intensity	• Labour-intensive; workers must physically collect samples, transport them to a lab, and perform testing	• Low labour intensity; sensors are automated, requiring only occasional maintenance, drastically reducing the need for manual labour
Data precision	• Results may vary based on the skill and consistency of personnel collecting samples	• High precision, with sensor-based data collection removing variability introduced by human factors, providing consistent and reliable data
Integration with smart systems	• Difficult to integrate into modern agricultural systems	• Easily integrates with other smart farming systems, such as automated irrigation or fertilization, and can be linked to predictive models for better resource management
Sustainability	• May contribute to soil disturbance and ecosystem disruption due to frequent sampling	• More sustainable, with minimal disturbance to the soil ecosystem as sensors are often left in place for long periods and do not require repeated physical interaction
Remote access and control	• No remote access; all data collection and analysis must be done in person	• Remote access allows for off-site monitoring and control through mobile apps or web platforms, making it convenient for users to manage large-scale or distant operations from anywhere

(*Continued*)

TABLE 3.1 (*Continued*) Traditional soil sampling methods versus IoT-enhanced soil sampling

DIMENSION	*TRADITIONAL SOIL SAMPLING METHODS*	*IOT-ENHANCED SOIL SAMPLING METHODS*
Automation capability	• Low to none; manual effort is required for the entire process	• High automation potential; the system can autonomously collect data, alert users to anomalies, and adjust field operations like irrigation or fertilization based on sensor inputs and algorithms
Long-term insights	• Difficult to gather long-term insights without frequent manual sampling and record-keeping	• Enables continuous monitoring over months or years, providing valuable long-term data on trends in soil health, which can be crucial for sustainable farming or environmental conservation
Response time to issues	• Slow; problems such as nutrient deficiencies or soil degradation might only be detected after significant time has passed	• Fast; immediate detection of issues like soil compaction, waterlogging, or nutrient imbalance allows for rapid interventions, preventing crop damage or soil deterioration
Suitability for large-scale operations	• Not ideal for large-scale operations due to high labour costs and time-intensive processes	• Well-suited for large-scale operations, as automation and real-time monitoring significantly reduce the need for manual intervention, making it easier to manage larger areas effectively
Examples of use	• Traditional soil testing for agriculture, environmental studies, and land reclamation projects	• Precision agriculture, smart farming, environmental monitoring, and automated irrigation systems that require continuous data on soil moisture, nutrient levels, and temperature

3.3.6 Smart Sensors for Soil Sampling and In Situ Monitoring

Other previous approaches are direct collection of samples from the ground using augers or shovels for more than one depth and position. These samples are then taken to a laboratory for physical property tests, such as moisture content, pH, texture, and content. While useful, it is very manual, time-consuming, and may result in delayed responses in a subsequent analysis of results. In addition, the use of arbitrary selection of the sample also takes a certain period of time; some fluctuations in the parameters of the soil may still be determined only in the subsequent sampling.

On the other hand, through the application of smart sensors, IoT enabled the reform of soil sampling through continuous and real-time collection of soil parameters. In other words, through IoT, it is possible to check on the soil's status at any one time just in case samples might not need to be taken frequently. Real-time monitoring provides more updated details on the condition and provides engineers and scientists ways to respond to the shifts in soil behaviour, including things like moisture pressure and instability.

Furthermore, systems based on IoT can collect information from different locations and viewpoints to study the properties of the ground in several points of interest. This also eliminates many logistic considerations, which are always associated with conventional approaches, and enhances the credibility of the collected data

$$S(t)=\sum_{i=1}^{N}\left[\alpha_i\cdot\frac{dM_i(t)}{dt}+\beta_i\cdot\frac{\int_0^T\left(\frac{\partial\Psi_i(t)}{\partial t}\cdot\Phi_i(t)\right)dt}{\log\left(1+\frac{\partial^2\theta_i(t)}{\partial t^2}\right)}+\gamma_i\cdot\left(\frac{\partial VWC_i(t)}{\partial t}\cdot\frac{\partial EC_i(t)}{\partial t}\right)\right]$$

$$M_i(t)=\frac{\int_0^T\frac{\partial\Omega_i(t)}{\partial t}\cdot\Psi_i(t)\,dt}{1+\log\left(1+\frac{\partial\Psi_i(t)}{\partial t}\right)}$$

$$\Phi_i(t)=\frac{\partial}{\partial t}\left(\frac{\partial\Psi_i(t)}{\partial t}+\lambda\cdot\frac{\partial^2\Psi_i(t)}{\partial t^2}\right)$$

$$\theta_i(t)=\frac{\int_0^T\Psi_i(t)\cdot\frac{\partial^2\Omega_i(t)}{\partial t^2}dt}{\log(1+VWC_i(t))}$$

$$\mathrm{VWC}_i(t)=\frac{\partial}{\partial t}\left(\frac{\partial\Omega_i(t)}{\partial t}\cdot EC_i(t)\right)$$

$$\mathrm{EC}_i(t)=\frac{d}{dt}\left(\frac{\partial\Psi_i(t)}{\partial t}+\mu\cdot\frac{\partial^2\Psi_i(t)}{\partial t^2}\right)$$

$$\frac{\partial S(t)}{\partial t}=\sum_{i=1}^{N}\left(\alpha_i\cdot\frac{\partial^2M_i(t)}{\partial t^2}+\beta_i\cdot\frac{\partial}{\partial t}\left(\frac{\partial\Psi_i(t)}{\partial t}\cdot\Phi_i(t)\right)+\gamma_i\cdot\frac{\partial^2}{\partial t^2}(VWC_i(t)\cdot EC_i(t))\right)$$

$$\frac{\partial M_i(t)}{\partial t}=\frac{\partial}{\partial t}\left(\frac{\int_0^T\frac{\partial\Omega_i(t)}{\partial t}\cdot\Psi_i(t)\,dt}{1+\log\left(1+\frac{\partial\Psi_i(t)}{\partial t}\right)}\right)$$

$$\frac{\partial\mathrm{VWC}_i(t)}{\partial t}=\frac{d}{dt}\left(\frac{\partial}{\partial t}\left(\frac{\partial\Omega_i(t)}{\partial t}\cdot EC_i(t)\right)\right)$$

Symbols Explanation:

- $S(t)$ is the soil stability or condition index at time t.
- $M_i(t)$ is the moisture content as detected by sensor i at time t.
- $\Psi_i(t)$ is the matric potential or soil suction for sensor i.
- $\Omega_i(t)$ is the water retention characteristic for sensor i at time t.
- $\Phi_i(t)$ is the flow potential related to the soil's water flow.
- $\theta_i(t)$ is the soil water content measured by sensor i at time t.
- $\text{VWC}_i(t)$ is the volumetric water content for sensor i at time t.
- $\text{EC}_i(t)$ is the electrical conductivity for sensor i at time t, related to soil salinity.
- $\alpha_i, \beta_i, \gamma_i$ are the weighting coefficients for different parameters.
- λ, μ are the regularization parameters for smoothness.
- N is the total number of smart sensors.

3.3.7 Measures for Internet of Things Device Constraints under Adverse Conditions

IoT devices face various environmental conditions when used in soil research, which is often complicated. Excessive moisture, the use of corrosive substances, and temperature fluctuations are especially damaging to soil sensors and all monitoring systems.

- **Extremes of Temperature**: Kinetic, pressure, and humidity sensors typically used in IoT may not be as accurate or reliable because they are susceptible to extended high or low temperatures, and soil conditions can be very different. Sensors are sensitive to temperatures, and extreme temperatures, such as low temperatures, can freeze or stop working while high temperatures can overwork them. Probes must be developed that can perform in these manners if eventual and precise measurements are to be obtained.
- **Moisture and Corrosion**: Water or soil moisture can damage or corrode the sensors. This is minimized by using corrosion-resistant materials in the control circuitry and ensuring that the sensors are housed in virtually leak-proof pouches. Maintenance and inspections are also required for these sensors to operate in such settings.
- Soil factors can also affect the performance of the sensor in some ways. Two of these types of soils are clay and sandy soils. For example, very high salinity or acidity may lead to corrosion or reduce the effectiveness of the sensor calibration. Perhaps, the difficulties could be bypassed by installing different sensors customized for various types of soil and conditions.
- **Access Control**: The data can, therefore, only be opened or manipulated by personnel who meet the strict access requirements. The use of RBAC and MFA can possibly enhance security and decrease the possibility of unauthorized access.

- **Regular Audits**: IoT can be tested using security audits and vulnerability assessments, which are routine procedures for identifying system vulnerabilities. Addressing these vulnerabilities as early as possible in the process might help avoid the sorts of breaches detailed above and maintain the integrity of the data moving forward.
- It also reveals that the data gathered through IoT can be cross-validated with the result of the traditional soil sampling and testing done in the laboratory. Soil governance structures that integrate data from the Internet of Things (IoT) with decision-support tools enhance real-time monitoring, predictive analytics, and sustainable land management practices. These systems may offer automated recommendations for soil management, monitoring, and intervention based on real-time data analysis.
- **Networks for Adaptive Monitoring**: As IoT devices become more widespread, so will their ability to adapt, allowing them to change their monitoring strategies in response to real-time data and outside influences. These networks will enable dynamic and responsive soil monitoring, optimizing data collection and improving the accuracy of soil assessments.

3.4 WIRELESS COMMUNICATION PROTOCOLS FOR SOIL DATA TRANSMISSION

Telematic communication is also relevant in IoT-based soil sampling in view of its suitability with the data transfer from isolated sensors to the monitoring systems. This has implied several parameters to be used to decide the sort of communication to be used such as transmission range, data rate, and power consumption conditions (Figure 3.1). Some of the most common wireless protocols used for soil data transmission include the following.

- **LoRaWAN (Long-Range Wide Area Network)**: Thanks to relay nodes which consume low power, LoRaWAN range can cover large areas for sending data like the state of the soil in an agriculture field or a large construction site. This is meant to be used where applications are operating at low data rates, and this makes it perfect for use when testing properties of soil that are not frequently altered.
- **Zigbee**: In particular, Zigbee is a type of short-range wireless protocol to be implemented in marginally powered devices of IoT. It is best suited for mesh networks where many sensors are worn, and they can transmit data to each other and then to the control point. Zigbee is used in relatively dispersed soil monitoring systems, such as smart farms or city construction sites.
- **NB-IoT (Narrowband IoT)**: NB-IoT is a cellular-based technology offering connectivity solutions designed particularly for IoT, low-power wide area

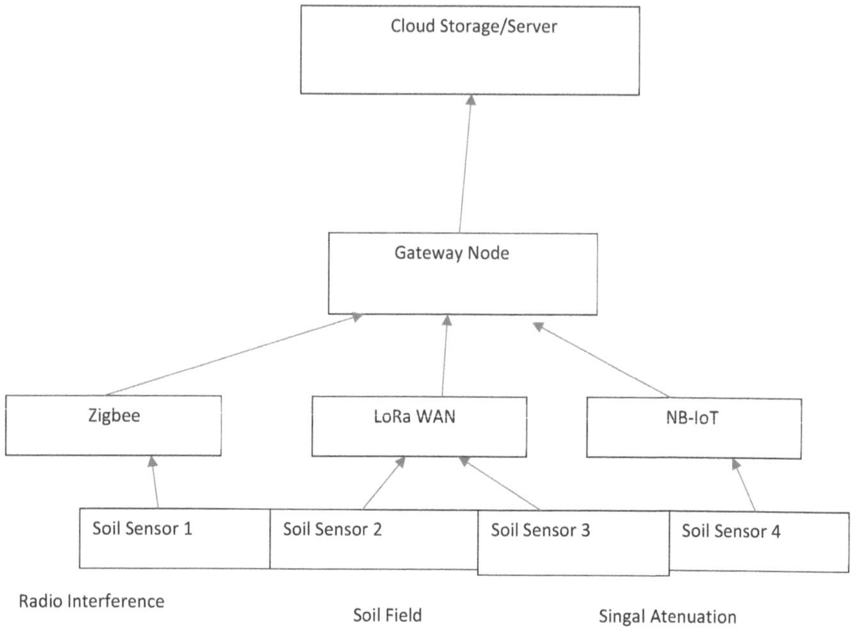

FIGURE 3.1 Wireless communication protocols for soil data transmission.

networks, and Low Power Wide Area Network (LPWA), particularly in areas where construction of other forms of communications structures is hard. By using NB-IoT, there is confidence in transferring data from soil sensors which have been installed in hard-to-access locations.

- **Wi-Fi**: Therefore, Wi-Fi applicants deliver high data rates with facilities that can be installed easily. They can be easily incorporated into the site facility arrangement. Nevertheless, it has rather high power consumption and radiates in a relatively limited area, which cannot be effectively utilized in long-term or large-scale projects.

3.5 STATISTICAL MODELS FOR SOIL DATA ANALYSIS

Statistical analysis of soil data is a fundamental aspect of geotechnical engineering. With the increasing integration of IoT in soil investigations, vast amounts of real-time data are collected, making statistical models essential for interpreting and deriving meaningful insights. These models help engineers and researchers understand soil behaviour, predict potential risks, and optimize construction projects. The following section outlines key statistical approaches, including descriptive statistics, predictive modelling, and applications of ML and AI in soil data interpretation.

3.5.1 Descriptive Statistics: Mean, Median, Mode, and Variance

Descriptive statistics makes up the initial level of analysis of data on soil by using it to transform and present huge amounts of data. In geotechnical investigations, soil properties like moisture content, density, and texture are often represented using basic statistical measures.

- **Mean**: A measure of the expected or typical performance of a soil because it will give an average value of this characteristic. For instance, the mean average moisture content of the soil at a site will determine how much extent the subsequently compacted or swelled or shrunk soil of the site can be.
- **Median**: The average of numbers arranged in order is considered to more properly represent the centre in data with a given condition. However, when there are large raw differences, that is, when the values are skewed and there are outliers in the data, the median is more meaningful than the mean.
- **Mode**: The mode which can provide information on typical and main characteristics of the given site's soil such as the main sort of the soil, the main degree of its moistening, and so on.
- **Variance**: Information about the variability of the data extent. Variance indicates how far specific properties of soil differ from the mean value. For example, high CVs of moisture content may indicate variability of the ground conditions that requires more investigation.

They can, therefore, be effectively used by geotechnical engineers to gain an early preliminary conclusion about the characteristics of a given soil before developing elaborate tests for the same.

3.5.2 Predictive Modelling for Soil Behaviour Forecasting

One of the common applications of geotechnical engineering is to forecast the manner in which the soil will respond to some conditions, and predictive analysis forms a valuable tool in this regard. This kind of analysis accepts the history of soil as input and then predicts the future results of this process which is very useful in risk assessment, project planning, or creation of various structures. Commonly, there are regression analysis data predictive models, time-series model, and geostatistical model.

- **Regression Analysis**: This technique assumes the co-variation of the soil characteristics and the other corresponding factors. For example, regression can forecast how the soil moisture content can change with temperature, or the rain/no rain information is valuable while designing the foundations or slope stability analysis.
- **Time-Series Forecasting**: This is among the longitudinal methods of soil property information because it pays more attention to the evaluation of trends and patterns of the soil for the future conditions expected by engineers. However, moisture together with temperature forms a good candidate for time-series models, whereby moisture affects the compaction of soils and their load--bearing capacity.
- **Geostatistical Models**: These models factor in space distribution of soil characteristics and are often coupled with geographical information technologies (GITs). This chapter demonstrates how geostatistical overview patterns in a site produce analytic models to reply to issues of worry touching on either vigour or moisture alterability of the soil.

Risk models enhance the decision-making for geotechnical activities as engineers can use predictors to anticipate activities that are likely to cause disruptions and then prevent them.

3.6 APPLICATIONS OF MACHINE LEARNING AND AI IN SOIL DATA INTERPRETATION

Geotechnical engineering is also growing fast, thanks to innovative technological advancements such as ML and AI in the analysis of soil information. When many decisions regarding IoT for soil monitoring are to be made, a lot of big data can be generated, and traditional statistical analysis tools can fail to handle these quantities. AI and ML are even better as the capability to perform exploration and pattern detection in data and, more so, in phrase prediction has improved.

- Supervised learning employs the use of early soil records to produce the other future values employing the learned model. For example, a supervised learning algorithm may be applied to calculate the probability of soil liquefaction in case of an earthquake by studying the soil behaviour in similar conditions.
- **Unsupervised Learning**: Finds and assigns "labels" to clusters and trends in the data in the same way. For instance, the clustering problem can be used to put in the same cluster similar soils with specific qualities with a view of gaining more understanding of the kind of soil that is available in a given area.

- **Reinforcement Learning**: Practice in a populated place with feedback to change algorithms which are used for certain tasks. Other real-life reinforcement learning models in geotechnical engineering may be modified and improved with new properties of soil as the project is performed, thus enhancing its predictions and decisions.
- **AI-Driven Soil Behaviour Forecasting**: Different models are capable of predicting how natural resources will respond in light of different circumstances based on the analysis of several aspects of soil, including moisture content, density, and temperature. They also allow for developing predictions regarding reactions of the soil to some factors, such as construction loads or weather conditions, which helps prevent potential threats and increase the effectiveness of construction operations.

3.7 IoT-ENABLED SOIL MOISTURE MONITORING AND PREDICTION

Soil moisture is an important parameter in geotechnical engineering practice mainly because it affects the strength, compaction, and stability of soils. With the use of IoT technologies, real-time monitoring of soil moisture becomes more accurate, efficient, and informative in enabling engineers to make decisions on construction and maintenance projects. This section will focus on the role of moisture in geotechnical technologies, analysis of tort, and methods of variability of moisture forecast with the help of IoT devices.

- Soil moisture is a crucial parameter in geotechnical investigations of a site due to its influence on the behaviour of the soil under loading, consolidation, expansion, swelling, shrinking, and its potential for liquefaction during an earthquake. For instance, changes in soil moisture may affect the bearing capacity of the soils on which the foundations sit, precipitating different types of settlement or other problems. When it comes to areas that are susceptible to flood or drought, for instance, monitoring of the soil moisture becomes extremely sensitive since floods or drought weaken the soil and make it less stable.
- In slope stability analysis, changes in moisture content because of rains or movement of water in the subsurface greatly influence the chances of a slide. Ture monitoring becomes more precise, effective, and informative, helping engineers make informed decisions during construction and maintenance projects. This section will cover the importance of soil moisture in geotechnical investigations, statistical analysis of moisture variations, and how IoT devices are changing the collection and forecasting of moisture data.

- **Importance of Soil Moisture in Geotechnical Studies**: Soil moisture content is an essential component of geotechnical studies because it affects the behaviour of the soil under load, its propensity for compaction, and the risk of issues such as swelling, shrinking, or even liquefaction during seismic events. For example, variations in soil moisture may alter the foundations' bearing capacity, leading to different types of settlement or other structural issues. In areas that are vulnerable to flooding or drought, soil moisture monitoring becomes even more crucial since extreme weather can weaken the soil and increase instability.
- In slope stability analysis, variations in soil moisture content due to precipitation or subsurface water movement can have a significant impact on the likelihood of landslides. To achieve safety and construct long-enduring geotechnical projects, structured decision-making requires timely detection of moisture contents in soil. Methods of statistics are employed in order to evaluate and compare changes in the amount of soil moisture at specific temporal and spatial instability. When IoT technologies are deployed, real-time soil moisture monitoring becomes more precise, effective, and informative, helping engineers make informed decisions during construction and maintenance projects. This section will cover the importance of soil moisture in geotechnical investigations, statistical analysis of moisture variations, and how IoT devices are changing the collection and forecasting of moisture data.
- **Importance of Soil Moisture in Geotechnical Studies**: Soil moisture content is an essential component of geotechnical studies because it affects the behaviour of the soil under load, its propensity for compaction, and the risk of issues such as swelling, shrinking, or even liquefaction during seismic events. For example, variations in soil moisture may alter the foundations' bearing capacity, leading to different types of settlement or other structural issues. In areas that are vulnerable to flooding or drought, soil moisture monitoring becomes even more crucial since extreme weather can weaken the soil and increase instability.
- In slope stability analysis, variations in soil moisture content due to precipitation or subsurface water movement can have a significant impact on the likelihood of landslides. In order to ensure safety and long-term stability in such geotechnical projects, proactive decision-making necessitates real-time monitoring of moisture levels.
- Statistical techniques are used to assess and analyse variations in soil moisture over time and at multiple sites. By applying these methods, engineers may be able to detect trends, patterns, and variations in the moisture content of the soil, with information on how this soil behaves in certain conditions.
- **Time-Series Analysis**: This technique is widely applied to analyse periodic variations in the volume of moisture in the soil, allowing, for instance, to define the increase in the moisture level typical for the wet season. This supports the engineer in planning when construction projects should take place and when they should integrate moisture control systems. The coefficient of variation of moisture data can be used to know the extent of variation of the

SM at a particular location, as well as seasonal patterns like rising moisture content in wet seasons. This aids engineers in scheduling construction projects and installing moisture control systems as necessary.

- The variance and standard deviation of moisture data can be analysed to determine the degree to which soil moisture levels vary at a given location. Low variance represents a steadier moisture level, while high variance is indicative of a fluctuating reality about the soil that requires closer monitoring.
- **Correlation and Regression Analysis**: The correlation and regression analysis methods are used to examine relationships between the soil moisture and other factors like temperature, precipitation, or type of soil. These investigations assist in anticipating subsequent moisture conditions and consider the ground surface's outdoor conditions. These statistical tools would enable engineers to have a deeper appreciation of how moisture fluctuations influence the response of the soil in geotechnical-related works. Smart appliances for measuring and forecasting moisture information have led to shifts in the collection, sharing, and analysis of soil moisture data through the use of IoT-enabled technologies, which enhance the identification of seasonal patterns like rising moisture content in wet seasons. This aids engineers in scheduling construction projects and installing moisture control systems as necessary.
- The variance and standard deviation of moisture data can be analysed to determine the degree to which soil moisture levels vary at a given location. Low variance denotes more constant moisture levels, whereas high variance can suggest unstable soil conditions that need more observation.
- **Correlation and Regression Analysis**: Correlations between soil moisture and other variables, including temperature, precipitation, or soil type, are studied using correlation and regression approaches. These investigations help predict future moisture levels and look at how the environment affects soil behaviour.
- Engineers can gain a better understanding of the effects of moisture variations and how they affect soil performance in geotechnical projects by employing these statistical methods.
- The collection, sharing, and analysis of soil moisture data have all changed as a result of IoT-enabled technologies. Other ways of determining soil moisture may take a longer duration and are less informative than the above imaging devices and include the use of samples manually taken or gravimetric evaluations. On the other hand, IoT devices offer a dynamic view of the soil states since they offer an opportunity to take continuous samples at frequent intervals.
- **IoT-Based Soil Moisture Sensors**: For instance, IoT-based soil moisture sensors include capacitive or resistive types; therefore, they are installed at different depths and positions within a site for real-time measurements of moisture content. These are connected by a wireless network to a principal

system which relays continual information to remote analysis servers in the cloud.

- **Data Collection and Transmission**: To send moisture data to distant servers, IoT devices employ wireless protocols like LoRaWAN, Zigbee, or cellular networks. Engineers can quickly respond to changes in moisture levels and remove any hazards by downloading and analysing this data in real-time.
- **Moisture Forecasting and Predictive Analytics**: IoT systems can generate accurate moisture forecasts by fusing real-time moisture data with historical patterns and environmental data. By using these projections, geotechnical engineers may better prepare and reduce risk by anticipating moisture fluctuations brought on by weather variations, shifting water tables, or construction activities.
- **Machine Learning for Moisture Prediction**: To uncover hidden trends and improve forecasting accuracy, sophisticated ML techniques can be used for moisture data provided by IoT. With the use of AI, these models can predict future moisture levels depending on a variety of factors, including soil composition, temperature, and rainfall.

3.7.1 Geospatial Data Integration and Mapping with IoT

How geotechnical engineers analyse and disseminate soil characteristics has change due to the intersection of the IoT model and geolocation services (Panigrahi et al 2023). Soil data obtained from IoT can be incorporated into geospatial analysis vi geographic information systems (GIS) and diverse enhanced geostatistical approaches yielding optimal and comprehensive patterns of how soil behaves over an extensiv region. This section reviews the application of geostatistics in spatial analysis, the linkage of IoT data with GIS maps, and visualization techniques in supporting soil propert analysis (Dado et al., 2020). GIS technologies help engineers to acquire, store, analyse and visualize data related to geography, including related spaces. IoT-generated soi data can be easily integrated into spatial analysis using GIS and advanced geostatistica techniques, providing more precise and thorough insights into the behaviour of soil ove wide areas (Oikonomidis et al., 2023).

This section discusses the use of geostatistics in spatial analysis, the integration o IoT data with GIS, and visualization tools to enhance understanding of soil properties GIS technologies allow engineers to gather, store, process, and present geographicall connected data. They do this by leveraging the data from IoT. Soil properties of a geotechnical site can be explained dynamically using geospatial and information technology tools in relation to sensor data acquisition (Panigrahi et al., 2023). An importan system of categorizing and bringing together data concerning the soil, in a geographica perspective, is GIS, which is enhanced through the use of IoT sensors for purposes o tracking parameters such as temperature, moisture, and density of soil. Soil informatio is constantly collected from IoT devices deployed at various locations and is then overlaid on geographic maps in GIS.

Information on soil quality as a physical environment is constantly provided by IoT sensors from different regions and overlaid on geographical maps in GIS (Dado et al., 2020). This, in a way, allows engineers to evaluate the variations of the soils in use in different applications and areas. Other environmental information such as topographical data, vegetation cover, and climate could also be mapped to assist in identifying other influences on soil response under GIS. The context being explored in this work is the application of IoT and GIS for enhancing decision-making based on a detailed analysis of soil parameters in geotechnical projects.

Geostatistical Evaluation of Soil Properties: Geostatistics, a branch of statistics dealing with spatial or spatiotemporal data, is an ideal tool for determining soil quality across various regions. For analysing trends, extrapolating soil behaviour in uninvestigated regions, and exploring the spatial variability of soil parameters, geostatistical techniques can be applied conveniently (Oikonomidis et al., 2023). Kriging is an extensively used method for spatial average estimation to interpolate the soil information that is sensed using IoT sensors. A full coverage is attained whereby the quality of the soil is inferred at unknown portions through the use of adjacent sensors.

Thus, when regionally variable soil conditions are identified, engineers can use geostatistical methods in combination with IoT to analyse patterns, risks, and propose optimal solutions for infrastructure projects. Visualization is an active part of decoding data derived from IoT because it transforms raw numerical data into something easily understandable. For the accurate representation of the soil conditions across a site, GIS, along with special software, is available to engineers (Panigrahi et al., 2023).

3.7.2 Case Studies: IoT-Driven Soil Investigations

They pointed out that this effort has led to great progress in geomechanics due to the integration of IoT technology. Experience has been gained in using real-time data acquisition and advanced analytical techniques in IoT-based soil research, which has advanced our understanding and management of soil-related challenges in most engineering undertakings. This section includes three case studies that show how IoT is being used in geotechnical applications on a practical level: investigating the stability of slopes, the options for construction foundations, and the hazard of unrestrained movement of soil.

3.7.3 Applications of IoT in Monitoring Slope Stability

Stability of slopes is a common issue in civil engineering practices, primarily in areas characterized by landslides and other types of soil movements. In particular, IoT has improved slope monitoring and evaluation for stability. Recently, IoT sensors were installed on a landslide-prone slope in the Swiss Alps in order to capture the variations of moisture, temperature, and pressure in real-time.

The sensor data was input to a central processing system which in turn tallied and scrutinized the data to identify early signs of instability. Clients were able to observe the variation of pressure and moisture distribution on the slopes' surface in real-time by applying GIS. There have been predictive models observed that indicated potential slope failure thus facilitated immediate intervention by drainage enhancements and soil stabilization. The following executive action helped enhance the safety of those living nearby, especially by significantly reducing the likelihood of landslides.

Adequate knowledge of soil characteristics is crucial when assessing engineering conditions for new constructions on challenging urban soils. One example is how IoT impacted the foundations when constructing a high-rise facility in Tokyo. To capture the behaviour of the soil during excavation and pouring, an IoT sensor network was developed.

To ensure that the soil conditions were higher than the defined design input, data of the pressures, densities, and moisture content accessible from IoT sensors were checked in real-time. Due to this real-time data, engineers could quickly change the foundation design if such behaviour of the soil is observed. The data, for instance, showed areas where there was more humidity than expected that would require additional drainages. This kinetic approach helped to provide structural support to the structure and foundation too.

Liquefaction is, therefore, the process that leads to the softening of normally consolidated saturated soil and its behaviour like a liquid during an earthquake. A new and unique project in the city of Christchurch, New Zealand, applies IoT to manage the probability of ground liquefaction in the seismic zone of a densely populated residential neighbourhood.

For soil density, pore water pressure, and moisture content variation identification in this experiment, IoT sensors were installed in the soil at different levels. Hence, risk assessment concerning soil liquefaction was obtained from the recorded information duly retrieved from the sensors during an earthquake. With this data, engineers would be able to define high-risk zones and then proceed to select more effective risk reduction methods using current geotechnical systems.

The vibro compaction and other additional measures of ground improvement were used only in some sites because of the higher pore water pressures than allowed, according to the data. It was also noted that IoT technology enhanced safety and infrastructure reliability by facilitating a better, faster response system to possible liquefaction issues.

3.8 CHALLENGES IN IoT-BASED SOIL DATA ANALYSIS

Geotechnical engineering has benefited from IoT applied to soil data and improved the decision-making process while providing real-time information. However, there are also some challenges in effectively employing this technology. Some of the main problems are data confidentiality and integrity, problems with anomalies and noises in the data, and the fact that IoT devices are often unable to meet performance demands.

- Measures for IoT device constraints under adverse conditions
- IoT devices face various environmental conditions when used in soil research, which are often complicated. Excessive moisture, the use of corrosive substances, and temperature fluctuations are especially damaging to soil sensors and all monitoring systems.
- **Extremes of Temperature**: Kinetic, pressure, and humidity sensors typically used in IoT may not be as accurate or reliable because of their susceptibility to extended high or low temperatures because soil conditions can be very different. Sensors are sensitive to temperatures, and extreme temperatures such as low temperatures can freeze or stop working while high temperatures can overwork them. Probes must be developed that can perform in these conditions if eventual and precise measurements are to be obtained.
- **Moisture and Corrosion**: Water or soil moisture can damage or corrode the sensors. This is minimized by using corrosion-resistant materials in the control circuitry and ensuring that the sensors are housed in virtually leak-proof pouches. Maintenance and inspections are also required for these sensors to operate in such settings.

Soil factors can also affect the performance of the sensor in some ways. Two of these types of soils are clay and sandy soils. For example, very high salinity or acidity may lead to corrosion or reduce the effectiveness of the sensor calibration. Perhaps, the difficulties could be bypassed by installing different sensors customized for various types of soil and conditions.

3.9 FUTURE TRENDS IN SOIL STATISTICS AND IoT FOR GEOTECHNICAL ENGINEERING

With the development of technology, there is potential for more enhanced use of IoT and soil statistics in geotechnical engineering. As a further evolution in soil investigations, promising significant breakthroughs in the field include IoT, data intelligence, and app innovations. This section examines prospective future technologies in IoT, new developments in ML and data analysis, as well as potential future findings that may influence soil monitoring.

3.10 CONCLUSION

In this chapter, we discussed the application of statistics and IoT solutions relating to geotechnical characterization. Of these, statistics of soil, which are among the basic forms of geotechnical data, provide important characteristics of the studied samples of soil as to their texture, density, and moisture. Statistical methods applied in data analysis, finally, help engineers in decision-making by determining the behaviour of soil and/or efficiency and/or capability of a construction idea, which is useful in the determination of civil projects' success and/or failure. In the discussion above, one can see that IoT has taken the means of improving soil assessments even one step ahead. Traditional methods of soil sampling also entail obtaining data that must then be transmitted physically to other centres that then process this data and release it; IoT has now made it possible to collect data through smart sensors which can then be transmitted wirelessly. The progress of soil monitoring is not solely helpful in the investigation but also essential for continuously assessing soil parameters such as stability analysis, soil moisture content, compaction, permeability, and overall soil health, which are crucial for agricultural, environmental, and civil engineering applications. By analysing all of the aforementioned reasons, it is understood that statistical models, including descriptive and predictive analysis, can be applied to the IoT-collected data. We transitioned to ML and AI possibilities for identifying the activity of the soil, using GIS to layer geographical information for improved understanding of the properties of the soil. Finally, examples from practice proved the importance of IoT application in geotechnical investigations and revealed that IoT-connected investigations of stability of slope, foundation, and risks of soil liquefaction are highly valuable in geotechnical practice. Still, some problems exist: challenging climate and data quality and integrity concerns. Soil statistics and the future for IoT are bright as progress will enhance quality, data, analysis, and forecasting.

REFERENCES

Ansarifar, J., Wang, L., & Archontoulis, S. V. (2021). An interaction regression model for crop yield prediction. *Scientific Reports*, *11*(1), 1–14.

Anupam, K., Swaroop, V., Lal, P. S., & Bist, V. (2015). Turning Leucaena leucocephala bark to biochar for soil application via statistical modelling and optimization technique. *Ecological Engineering*, *82*, 26–39.

Borrelli, P., Robinson, D. A., Fleischer, L. R., Lugato, E., Ballabio, C., Alewell, C., & Panagos, P. (2017). An assessment of the global impact of 21st century land use change on soil erosion. *Nature Communications*, *8*(1), 1–13.

Burdett, H., & Wellen, C. (2022). Statistical and machine learning methods for crop yield prediction in the context of precision agriculture. *Precision Agriculture*, *23*(5), 1553–1574.

)ado, W. T., Deines, J. M., Patel, R., Liang, S. Z., & Lobell, D. B. (2020). High-resolution soybean yield mapping across the US Midwest using subfield harvester data. *Remote Sensing*, *12*(21), 3471.

)obos, E., Montanarella, L., Nègre, T., & Micheli, E. (2001). A regional scale soil mapping approach using integrated AVHRR and DEM data. *International Journal of Applied Earth Observation and Geoinformation*, *3*(1), 30–42.

Ielber, P., Bischke, B., Habelitz, P., Sanchez, C., Pathak, D., Miranda, M., & Dengel, A. (2023). Crop yield prediction: An operational approach to crop yield modelling on field and subfield level with machine learning models. In *IGARSS, IEEE International Geoscience and Remote Sensing Symposium* (pp. 2763–2766). Pasadena, California.

effries, G. R., Griffin, T. S., Fleisher, D. H., Naumova, E. N., Koch, M., & Wardlow, B. D. (2020). Mapping sub-field maize yields in Nebraska, USA by combining remote sensing imagery, crop simulation models, and machine learning. *Precision Agriculture*, *21*, 678–694.

ohnson-Bailey, J., Valentine, T., Cervero, R. M., & Bowles, T. A. (2009). Rooted in the soil: The social experiences of Black graduate students at a southern research university. *The Journal of Higher Education*, *80*(2), 178–203.

Ma, Y., Liang, S. Z., Myers, D. B., Swatantran, A., & Lobell, D. B. (2024). Subfield-level crop yield mapping without ground truth data: A scale transfer framework. *Remote Sensing of Environment*, *315*, 114427.

Najjar, H., Miranda, M., Nuske, M., Roscher, R., & Dengel, A. (2024). Explainability of sub-field level crop yield prediction using remote sensing. *Xiv preprint arXiv*:2407.08274.

Oikonomidis, A., Catal, C., & Kassahun, A. (2023). Deep learning for crop yield prediction: A systematic literature review. *New Zealand Journal of Crop and Horticultural Science*, *51*(1), 1–26.

Panigrahi, B., Kathala, K. C. R., & Sujatha, M. (2023). A machine learning-based comparative approach to predict the crop yield using supervised learning with regression models. *Procedia Computer Science*, *218*, 2684–2693.

Pathak, D., Miranda, M., Mena, F., Sanchez, C., Helber, P., Bischke, B., & Dengel, A. (2023). Predicting crop yield with machine learning: An extensive analysis of input modalities and models on a field and sub-field level. In *IGARSS, IEEE International Geoscience and Remote Sensing Symposium* (pp. 2767–2770). Pasadena, California.

Sensors in Geotechnical Engineering for Structural Physical Condition Monitoring

4

4.1 INTRODUCTION TO IoT IN STRUCTURAL CONDITION MONITORING

The relevance of structural health monitoring (SHM) is immeasurable, especially i the field of geotechnical engineering. Thanks to real-time analysis of structural an soil conditions, engineers can monitor the health of infrastructure on an ongoing basis making it possible to spot potential problems early before they become major issue (Hajovsky & Pies, 2012). This proactive method serves to secure infrastructure reli ability and extend the life of essential structures, thereby saving time and resources Internet of Things (IoT) sensors are instrumental in evolving monitoring methodolo gies due to the magnitude of real-time data that they generate. By embedding sensor in soil and structural elements, continuous information is delivered regarding severa parameters, such as the water content in the ground or the amount by which the eart has moved (Hérail & Bit-Monnot, 2023). This insight equips engineers to take decisiv

DOI: 10.1201/9781003537625-4

action, respond to changing circumstances, and implement corrective actions promptly for risk management.

Today, we see a broad range of IoT sensors—often quite specialized. Between accelerometers that can measure vibration and strain gauges that detect deformation, various sensors allow for comprehensive monitoring of soil conditions. This data can include temperature and humidity sensors, which enhance understanding of environmental effects, as well as pressure sensors which provide essential load distribution knowledge (Ma et al., 2015). No single type of sensor can provide a complete picture of soil behaviour, though all types have their strengths and use cases, enabling previously unattainable advancements in geotechnical engineering.

4.1.1 Importance of Structural Condition Monitoring

The run-flat sensors are critical, and the installation method that will ensure they are correctly positioned needs the best strategy to be used. This is key to how the data will be used, so when we talk about deciding where you put your sensors and whether it should be a distributed network or a more localized one makes all the difference. The better the sensors are placed, the clearer picture the engineers will get from a soil type and come to more correct conclusions of how stable a building needs to be.

An essential component used in monitoring through IoT is real-time data acquisition systems which enable the collection and transmission of data. They are meant for processing vast amounts of information and delivering quick analytics, so the engineers can act as quickly as they need to when specific challenges crop up. Complementing this information, data fusion techniques provided additional data from various sources, allowing a more enriched assessment of the state of the soil.

The use of IoT sensors as part of SHM systems is a major development in geotechnical engineering. In this manner, the integration of IoT within SHM can be optimized by means of creating frameworks that facilitate the interworking of the sensors and SHM platforms. This integration also helps in one-of-a-kind observations of the structural performance in real-time mode and ensures that the indicators that serve as a basis for creating maintenance and risk management plans are precise.

4.1.2 Types of IoT Sensors for Structural Monitoring

Accelerometers: Vibration and tilt are two important measurements which are commonly used in the evaluation of structures, both natural and man-made.

Accelerometers are one of the primary sensors needed for identifying vibrations and tilt in structure systems (Figure 4.1). These devices determine acceleration loadings within a structure, which can allude to oscillations and shifts or other forms of structural instability. In soil monitoring, accelerometers are used to determine the effects of ground vibrations or seismic activities on various structural elements, helping to determine the stability and safety required for their use.

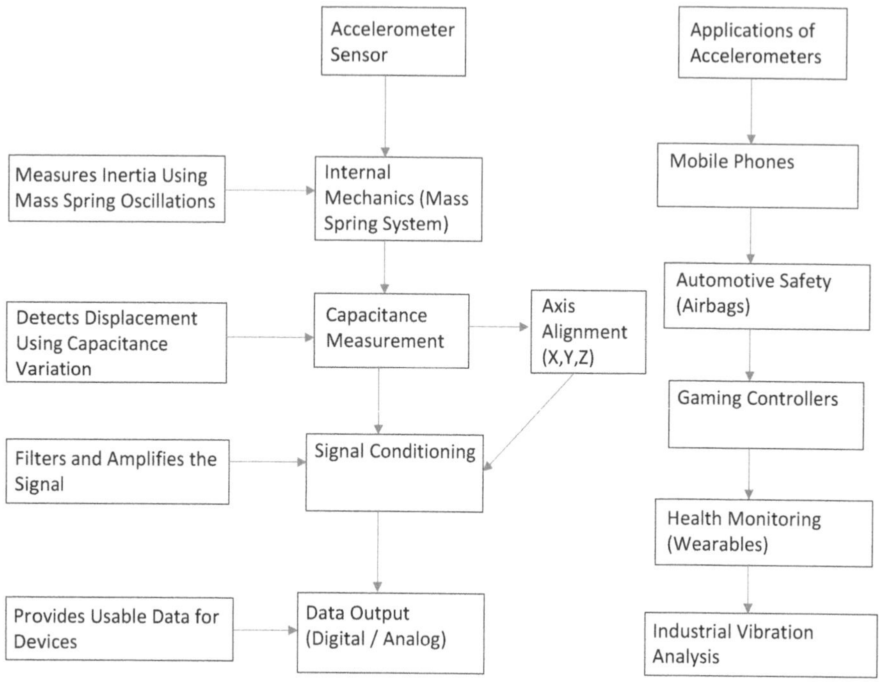

FIGURE 4.1 Accelerometers.

Strain Gauges: Measuring and tracking of structural deformations

There are other tools used in measuring the deformation of structures, and one of them is the strain gauge in Figure 4.2. These sensors quantify the level of deformation which the material is subjected to by forces being applied to it. The utilization of strain gauges provides engineers with opportunities to measure stress and deformation in specific areas of a structure and predict which aspects of a structure might benefit from added support or reinforcement.

Temperature and Humidity Sensors:

Temperature and humidity probes present worthwhile data about the climate affecting the soil as well as the structures in Figure 4.3. Soil is a very sensitive medium and responds to changes in temperature and moisture content by varying its load-bearing capacity and compacting ability. If these parameters are frequently observed, engineers can get a clearer perspective of ways in which environmental shifts affect soil response and respond accordingly.

Pressure Sensors: Load Monitoring:

Load monitoring is the main application of pressure sensors, given the ability to gauge the force being applied on structures or the soil in Figure 4.4. These sensors are specially required when evaluating how loads applied on a structure are segregated: an aspect critical in determining the overall stability of the structure. Pressure sensors are used in geotechnical settings to measure changes of pressure in soil under varying loads, thereby giving directions on likely implications of soil on structural behaviour.

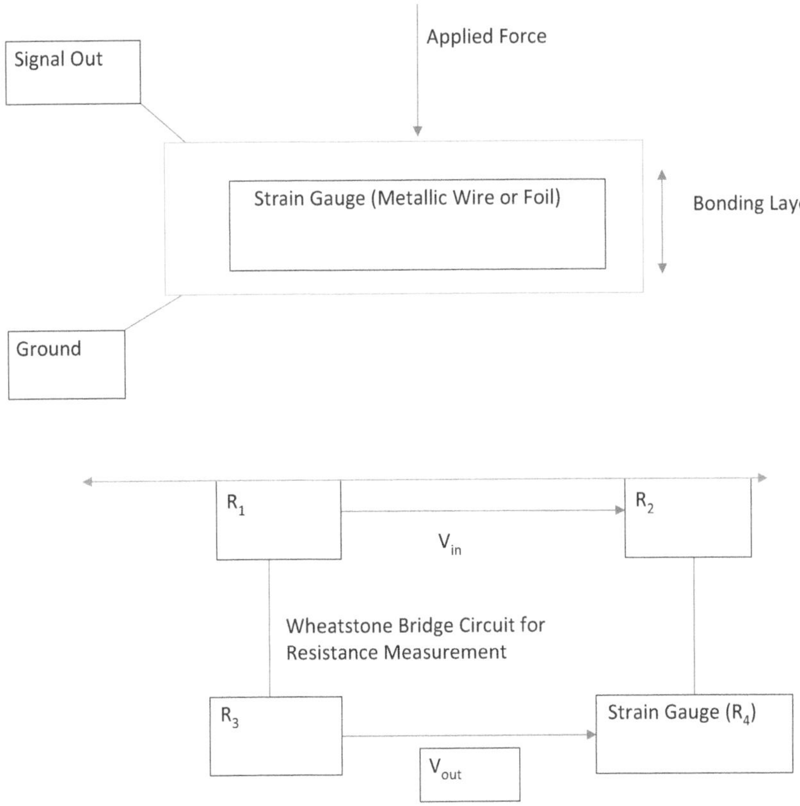

FIGURE 4.2 Strain gauges.

4.1.3 Role of IoT Sensors in Enhancing Monitoring Techniques

Data analytics is another crucial gadget that helps analyse lots of data gathered by IoT sensors. The data found in the 3D model can then be used with machine learning (ML) and artificial intelligence (AI) algorithms to identify possible problems and plan the appropriate time for maintenance. Further support is provided by the availability of the outcome to be delivered to cloud technologies, permitting for real-time monitoring and evaluation of the state of the soil from virtually any location.

There is a need to identify the appropriate communication standards needed for IoT sensors to function optimally in geotechnical operations. LoRa, Zigbee, and 5G technologies make data transfer reliable and energy-efficient, ensuring that information is delivered at the right time with high security and minimal latency, thereby enhancing communication in IoT and smart infrastructure applications. Security concerns that help sustain the integrity of the network system work to ensure that no unauthorized persons can access restricted areas, thus making the monitoring system more reliable.

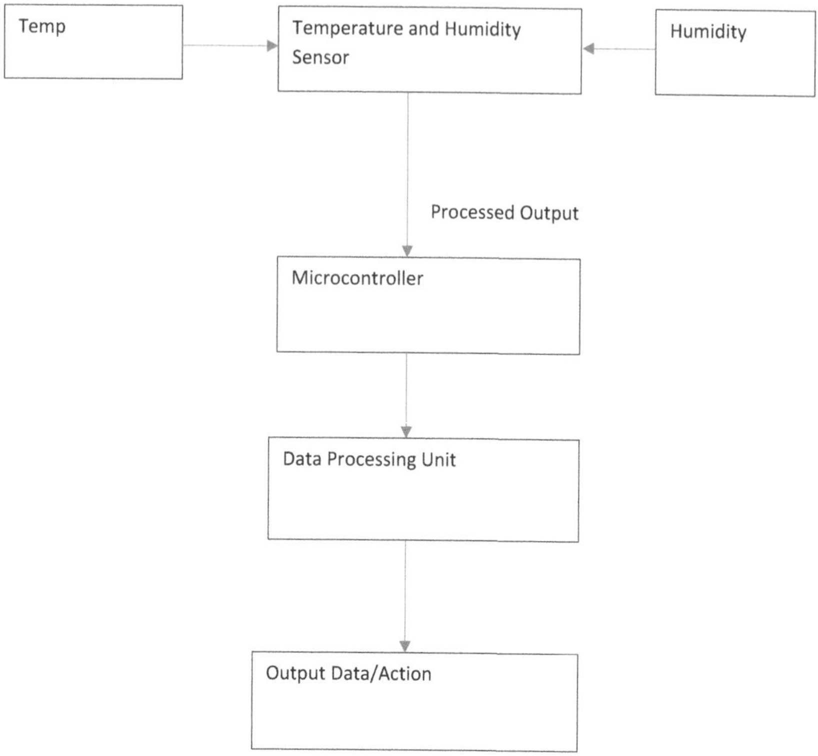

FIGURE 4.3 Temperature and humidity sensors.

An increasingly important element of IoT-enabled soil monitoring is edge computing. Edge devices allow quicker decision-making by reducing latency and processing data immediately on-site. This feature is especially helpful in dynamic settings where real-time analysis is essential. Case studies that show how edge computing improves the efficacy and efficiency of geotechnical monitoring include those that deal with smart infrastructure (Table 4.1).

4.1.4 Real-Time Monitoring Considerations

This is one among the fundamental aspects of state-of-the-art IoT-based sensor systems. Real-time monitoring must take into account a few things to be successful. The transmission of all data from sensors to central systems for processing needs to be reliable and timely. This often involves selecting appropriate wireless communication protocols and ensuring network reliability. Processing data efficiently requires scalable systems that can move a large amount of data and analyse it before it goes stale. Since the sensors will work continuously and will not accept any maintenance or repairing, power supply and battery life durations also come into play. By considering these factors, you

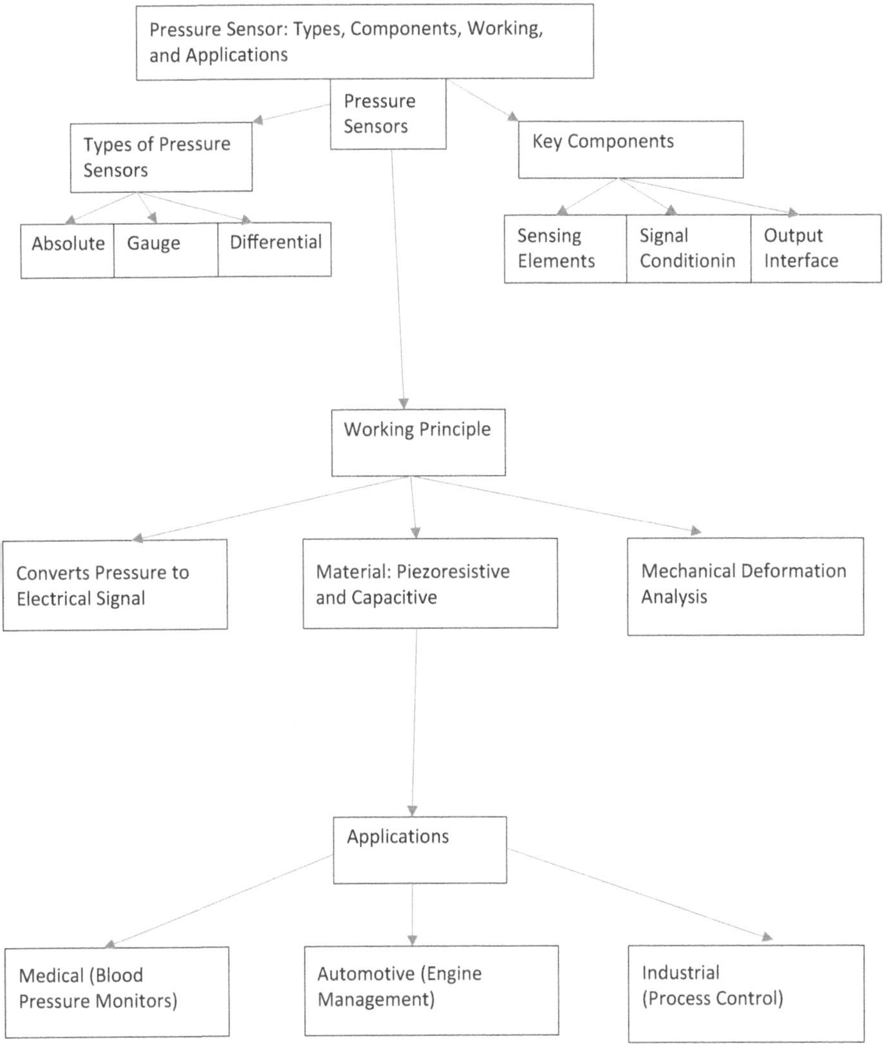

FIGURE 4.4 Pressure sensors.

are sure that the monitoring system will be able to give alerts and updates in real-time and a fast fix for any new issues (Hassani et al., 2021).

4.1.5 Architectural Framework of IoT-Based Monitoring Systems

The IoT-based monitoring system design is made in a manner that the collection, transmitting, and analysing of the data are made easier. Normally, the system has several

TABLE 4.1 IoT sensors in enhancing monitoring techniques

FIELD	*DESCRIPTION*	*SENSOR TYPES*	*APPLICATIONS*	*KEY BENEFITS*
Environmental monitoring	IoT sensors collect data on environmental conditions such as temperature, humidity, air quality, and pollution levels, enabling real-time insights for management	Temperature sensors, humidity sensors, gas sensors, pressure sensors	Smart cities, agriculture, pollution control, wildlife conservation, disaster management	Early detection of hazards, better resource management, real-time data collection, and improved decision-making
Health monitoring	IoT sensors track patient vitals such as heart rate, glucose levels, blood pressure, and physical activity, enhancing remote and continuous health surveillance	Wearable sensors, biosensors, ECG/EKG sensors, glucose monitors	Remote patient monitoring, fitness tracking, chronic disease management, emergency response	Increased access to healthcare, early diagnosis, continuous monitoring, improved patient outcomes
Industrial monitoring	IoT sensors monitor machinery performance, vibration, temperature, and pressure in industrial settings, allowing predictive maintenance and fault detection	Vibration sensors, temperature sensors, pressure sensors, proximity sensors	Manufacturing, oil and gas, energy, transportation, smart factories	Reduced downtime, increased equipment lifespan, cost savings, enhanced operational efficiency
Infrastructure monitoring	IoT sensors provide real-time monitoring of structural integrity, load-bearing capacity, and other metrics for infrastructure such as bridges, buildings, and roads	Strain gauges, accelerometers, pressure sensors, vibration sensors	Smart cities, transportation, building management, civil engineering	Improved public safety, early warning systems for structural failures, efficient resource allocation for maintenance
Agricultural monitoring	IoT sensors monitor soil moisture, nutrient levels, temperature, and crop health to optimize irrigation, fertilization, and overall farm management	Soil moisture sensors, pH sensors, temperature sensors, light sensors	Precision agriculture, smart farming, greenhouses	Optimized resource use, increased crop yield, reduced waste, better environmental sustainability, real-time adjustments

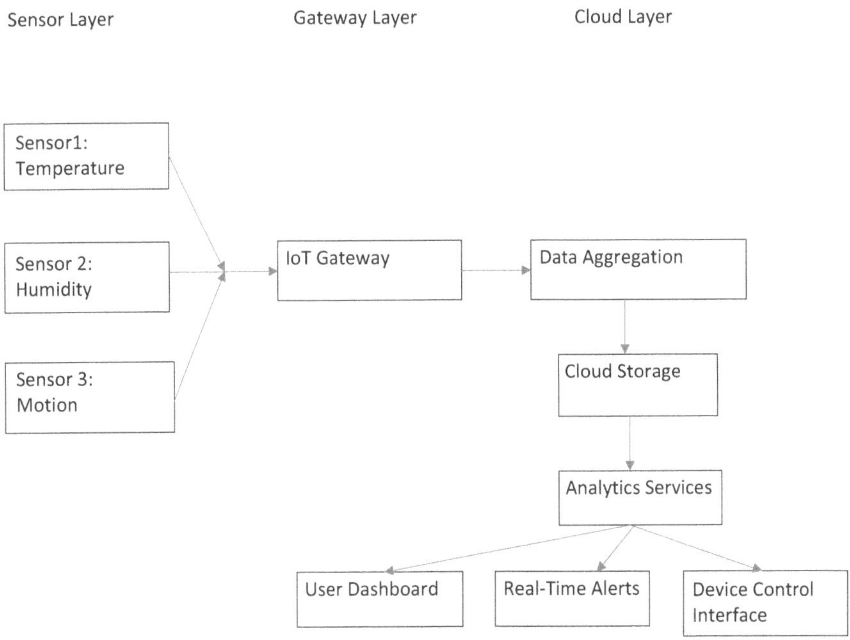

FIGURE 4.5 Architectural framework of IoT-based monitoring systems.

levels: the first one includes the sensors collecting information from the environment. These sensors are interfaced with data acquisition devices that acquire and process the raw data. The third layer is the communication networks through which the preprocessed data is sent to the central system. These networks can include wireless technology such as LoRa, Zigbee, or 5G, depending on the monitoring site that is shown in Figure 4.5. Data management systems at the central level refer to the process of data collection and warehousing at the central level while analysis of these data is made by reference to platforms and algorithms developed and implemented at the central level. This multilayered architecture guarantees that data processed through the sensors will be channelled appropriately in the system to facilitate better decision-making (Mehrabi et al., 2021).

4.2 WIRELESS SENSOR NETWORKS (WSNs) FOR STRUCTURAL HEALTH

Wireless sensor networks (WSNs) play a crucial role in collecting and forwarding data from several sensors placed at a given monitoring location. WSNs allow continuous and simultaneous monitoring of structural and soil states through the integration of a series of sensor nodes, which simplifies the compilation of data from different components in the monitored environment, providing a broad view and improving the efficiency of the monitoring system (Singh et al., 2023).

The effectiveness of multiple sensors is comprehensive in translating the dynamics of soil and structures. Many devices commonly used include accelerometers, strain gauges, temperature and humidity sensors, pressure sensors, and WSNs, each providing different pieces of information for the monitoring process (Moreu et al., 2015).

With these IoT sensors, engineers can devise early interventions to counter potential problems, ensuring that structural failures and infrastructure degradation are avoided through real-time data collection (Sadhu et al., 2023). The incorporation of sensors in a monitoring system increases its efficiency, making damage assessment and decision-making more precise (Singh et al., 2023). This integrated approach is essential, given the complex nature of issues in geotechnical engineering and structural monitoring.

As IoT technology advances, more developed sensors will enhance monitoring, improving our ability to predict and control soil and structural data, ultimately increasing the performance and safety of built structures (Sadhu et al., 2023).

4.3 SENSOR PLACEMENT STRATEGIES

In implementations of IoT sensors, the locations of the sensors are important in providing the right data of geo-structural performance. Proper placement enables the sensors to pick the relevant information on the state of the soil and structures to enable the right decision to be made and at the correct times. This section looks at some major steps of sensors' location: best places, networks for their connection, and real-time monitoring.

Of these techniques, choosing the right locations to install sensors is elemental to successful monitoring. The locations are usually defined with reference to the points in the structures that are most sensitive, or the soil layers most likely to undergo changes. For instance, in a building application, the sensors could be attached to the load-bearing walls, the foundation, and areas of high stress concentration. Also, in soil monitoring, an instrument is installed at various depths and near significant components such as retaining walls or embankments. When engineers choose those sites, the most appropriate data is collected to determine the health and stability of a particular structure or soil.

4.3.1 Distributed versus Centralized Sensor Networks

Thus, the potential distinction between the distributed and centralized sensor networks remarkably defines the performance of the monitoring system. Distributed sensor networks are characterized by the deployment of many sensors at different sites in the monitoring region, thereby forming a network of sensors. It enables examination from a spatial perspective, as many situations require a finer resolution to detect more localized issues. On the other hand, a centralized sensor network is characterized by a few sensors mounted in strategic places and a central hub that competently compiles and interprets the data. Although this monitoring approach will ease the management of data as well as the overall costs, it might lower the spatial density of monitoring. Thus, the decision about choosing one of these configurations highly depends on the characteristics of the particular monitoring project: its size and complexity, as well as the available funds (Table 4.2).

TABLE 4.2 Distributed versus centralized sensor networks

CRITERIA	*CENTRALIZED SENSOR NETWORKS*	*DISTRIBUTED SENSOR NETWORKS*
Network structure	A central hub or server collects data from all sensors and processes it in a centralized location	Each sensor node processes data locally or collaboratively with neighbouring nodes, distributing the processing across the network
Data processing	Data is transmitted to a central system where all data aggregation, processing, and decision-making occur	Data is processed at the sensor level or shared between nodes for distributed data processing, reducing the load on a central server
Scalability	Limited scalability due to reliance on a central server, which can become a bottleneck as the network grows	Highly scalable, as processing is distributed across nodes, allowing the network to grow without overloading a single central unit
Fault tolerance	Less fault-tolerant; if the central hub fails, the entire network can be compromised or lose functionality	More fault-tolerant, as the failure of one node does not affect the entire system. Data can still be processed by other nodes in the network
Energy efficiency	Centralized systems typically require higher energy consumption due to constant communication between sensors and the central hub	More energy efficient, as data processing is distributed, reducing the need for long-distance communication with a central system
Latency	Higher latency due to the time required to transmit data to the central hub and then process it	Lower latency since data processing can happen closer to the source, resulting in faster response times
Communication overhead	High communication overhead, as raw data from multiple sensors must be continuously sent to the central hub for processing	Lower communication overhead, as nodes process data locally or within small clusters, minimizing the need for frequent communication

(Continued)

TABLE 4.2 (*Continued*) Distributed versus centralized sensor networks

CRITERIA	*CENTRALIZED SENSOR NETWORKS*	*DISTRIBUTED SENSOR NETWORKS*
Network complexity	Simple network setup, as sensors only need to communicate with the central hub	More complex network setup, as each node requires the ability to communicate with multiple other nodes and handle processing tasks
Data security	Centralized control allows for simpler implementation of security protocols but creates a single point of failure that can be a security risk	More secure as data is distributed, reducing the risk of a single point of failure; however, more complex security protocols are needed
Cost	Can be more cost-effective for smaller networks but becomes expensive to scale due to the need for powerful central servers	Lower costs for large-scale deployments as there's no need for a powerful central server; however, individual nodes may be more complex
Maintenance	Easier to maintain as all updates and changes are made at the central hub	More challenging to maintain as each node may require individual attention and updates
Use cases	Best for small to medium-sized networks where data volume is manageable and central processing is sufficient	Ideal for large, complex networks like smart cities, environmental monitoring, or large-scale IoT deployments requiring real-time data

4.3.2 IoT Sensor Data Acquisition Systems

In the context of geotechnical engineering, IoT sensor data acquisition systems have a significant function in the data capture process whereby they identify data acquisition and processing for the various sensors installed in the monitoring sites. The intended uses of these systems are to develop precise and up-to-date information about the soil and the structure so that appropriate management and action can be taken. In this chapter, we review the major subcategories of IoT-based sensor data acquisition systems and describe their architecture, real-time data processing, and data integration (Dackermann et al., 2014).

4.3.3 Real-Time Acquisition and Delivery

Real-time data collection and transmission are important aspects of IoT, hence the proposed system. Measuring involves the steady and consecutive acquisition of readings from instruments which may include soil moisture content, strains on structures, or even other climacteric conditions. For efficient processing, this data has to be sent and hared with central systems for further analysis quickly. Telecommunication uses wireess communication channels, and these must be able to support data traffic by a large umber of sensors. Common features, which facilitate proper time delivery of data and rovide for data flow, include data buffering and prioritization to control or prevent data verload within the networks. Transmitting data in real-time enables quick analysis and eaction, which is crucial for identifying and resolving problems as they emerge.

4.3.4 Subtopic Sensing Sensor Calibration and Drift Issues

- **Calibration Accuracy:** Sensors need to be calibrated correctly to help capture measurements that get the desired results. Long periods eventually contribute to structural fatigue, material degradation, and overall system inefficiency. Delamination causes may include errors in calibration, which in turn generate inaccurate data for structural health ratings.
- **Regular Calibration Requirements:** However, due to the difference in the environments the system is subjected to, sensors for control are usually calibrated periodically to reduce drift. A fascinating and sophisticated process of modifying the characteristics of the sensors is found in the second step. There should be set schedules for inspections and calibration in order to maintain high accuracy in the future.
- **Calibration Procedures and Tools:** Tools and procedures in calibration is a way of making it effective. This means getting familiar with the calibration type and lags related to every kind of sensor integrated into the supply chain. Desensitization of personnel to calibration procedures is also crucial when it comes to regulating the sensors.

4.3.5 Functions of Sensors in Identifying Structural Imperfections

In its current form, IoT sensors are crucial in identifying poor structural work before it leads to major issues that could prove disastrous. These strategically placed sensors keep an eye on several parameters in crucial infrastructures:

- **Monitoring Structural Health:** Some of the usual structures which are observed for changes, stress, or vibration are buildings, towers, bridges, and dams. Three types of instruments are employed for this: the strain gauge, the accelerometer, and the displacement metre. Slight variation in these investments could be a sign that there are issues or perhaps structural failures.
- **Real-Time Data Analysis:** These sensors collect information that is then sent to central systems in real-time to allow complex algorithms to address critical decision-making, optimize performance, and enhance system

efficiency. This process helps determine any deviations or substandard structural defects before they develop into major problems.

- **Precautionary Measures:** Structural problem detection from an early time ensures that corrective measures are taken before huge tragedies or total loss of property occurs. For example, an automatic alarm system would detect a steep variation of certain structural characteristics, which would need quick exposure and rectification.

4.4 METHODS OF COMBINING DATA TO GAIN BETTER UNDERSTANDING

Data fusion techniques are used for augmenting the data collected by IoT sensors, thu increasing its value. By consolidating data from many sources, these techniques offe a more comprehensive view of the monitored environment. Data fusion is simply con solidating data from a variety of sensors to create one dataset that is able to present bot the physical health of the structure and soil condition. Integration of data provided b multiple sensors can improve accuracy and mitigate the impact of measurement error through techniques such as statistical processing and sensor fusion. Thereby, advance analytics—including ML algorithms—further enhance data fusion by identifying pat terns, correlations, and anomalies that are not visible from individual sensor readings A better knowledge allows us to handle structural and geotechnical systems more skil fully and take wise decisions.

4.5 INTEGRATION OF IoT SENSORS WITH STRUCTURAL HEALTH MONITORING (SHM) SYSTEMS

Geotechnical engineering is moving towards the use of IoT sensors/SHM, and these combinations are some of the outstanding geotechnical engineering advancements tha probably detect infrastructure systems in a much better manner. By making use of IoT technology and SHM systems, this integration provides a complete assessment systen to monitor the structural health state. In this review, the constituent elements for inte grating IoT sensors in SHM systems are reviewed as integration frameworks, interoper ability, and case of bridge monitoring.

$$S(t)=\sum_{i=1}^{N}\left(\alpha_i\cdot\int_0^T\left[\frac{\partial^2\sigma_i(x,t)}{\partial x^2}+\beta_i\cdot\frac{\partial\varepsilon_i(t)}{\partial t}\right]dx\right)+\gamma_i\cdot\int_0^T\frac{\partial^3\theta_i(x,t)}{\partial x^3}\cdot W(t)\,dx$$

$$\varepsilon_i(t)=\frac{1}{T}\int_0^T\left(\frac{\partial u_i(t)}{\partial t}\right)dt+\lambda\cdot\frac{1}{T}\int_0^T\left(\frac{\partial^2 u_i(t)}{\partial t^2}\right)dt$$

$$\sigma_i(x,t)=\frac{E_i}{1-\nu^2}\cdot\left(\frac{\partial u_i(x,t)}{\partial x}+\nu\cdot\frac{\partial v_i(x,t)}{\partial y}\right)$$

$$\theta_i(x,t)=\frac{M_i(x,t)}{E_i\cdot I_i}$$

$$W(t)=\frac{1}{T}\int_0^T\log\left(1+\frac{\partial\eta_i(t)}{\partial t}\right)dt$$

$$M_i(x,t)=\int_0^L\left(\frac{\partial^2 u_i(x,t)}{\partial x^2}+\mu\cdot\frac{\partial^3 u_i(x,t)}{\partial x^3}\right)dx$$

$$\frac{\partial S(t)}{\partial t}=\sum_{i=1}^{N}\left(\alpha_i\cdot\frac{\partial}{\partial t}\left(\int_0^T\left[\frac{\partial^2\sigma_i(x,t)}{\partial x^2}+\beta_i\cdot\frac{\partial\varepsilon_i(t)}{\partial t}\right]dx\right)\right)$$

$$+\sum_{i=1}^{N}\gamma_i\cdot\frac{\partial}{\partial t}\left(\int_0^T\frac{\partial^3\theta_i(x,t)}{\partial x^3}\cdot W(t)\,dx\right)$$

$$=\sum_{i=1}^{N}\left(\alpha_i\cdot\int_0^T\frac{\partial^3\sigma_i(x,t)}{\partial x^2\,\partial t}dx+\gamma_i\cdot\int_0^T\frac{\partial^4\theta_i(x,t)}{\partial x^3\,\partial t}\cdot W(t)\,dx\right)$$

Symbols Explanation:

$S(t)$ is the structural health indicator at time t.
$\sigma_i(x,t)$ is the stress at point x and time t in sensor i.
$\varepsilon_i(t)$ is the strain measured by sensor i at time t.
$\theta_i(x,t)$ is the angular displacement or curvature at point x and time t for sensor i.
$W(t)$ is the weighting function dependent on time t.
$M_i(x,t)$ is the moment in the structure monitored by sensor i.
$u_i(x,t), v_i(x,t)$ are the displacement components in the x and y directions for sensor i.
E_i is the Young's modulus of the material for sensor i.
ν is the Poisson's ratio of the material.
I_i is the moment of inertia of the structure for sensor i.

$\alpha_i, \beta_i, \gamma_i$ are the weighting factors for stress, strain, and curvature, respectively.
λ, μ are the regularization parameters.
N is the number of IoT sensors.
L is the length of the structural member being monitored.
T is the observation period or total time considered.

4.5.1 Integration of IoT and SHM: Two Different Frameworks

For IoT sensors to work together with SHM systems, strong integration formats have to be employed successfully. These frameworks define the procedures and interfaces that allow the integration of sensors, SHM systems, and their respective data streams. A typical integration framework includes several components:

- **Sensor Data Acquisition:** This is done through gathering the information obtained from the IoT sensors installed in the structure. Parameters could include strain, vibration, temperature, as well as pressure sensors.
- **Data Transmission and Communication:** Data acquired through the process is relayed to a central system by use of communication networks. This step guarantees rational accumulation and transmission of data obtained from various sensors in a real-time manner.
- **Data Integration and Analysis:** The transmitted data is used in the SHM system, whereby the data informs the health conditions of the structure. The collected data goes through a number of analytical escape, and algorithms then analyse the data to enumerate problems, maintenance requirements, and recommendations.
- **Visualization and Reporting:** This data is later presented in the form of dashboards and reports where engineers/decision-makers can understand the data and take the required action.

4.5.2 Interoperability between Sensors and SHM Platforms

The issue of interoperability plays a major role when choosing IoT sensors to complement SHM systems. It means the compatibility of multiple sensors and the platforms where data can be exchanged without any complication. Achieving interoperability involves several key considerations:

- **Standardized Protocols:** They allow sensors and SHM platforms to exchange data in an orderly way. MQTT, HTTP, and CoAP are protocols, for example, that are often utilized in IoT systems for communication between devices.

- **Data Formats and Integration:** For integration to be seamless, data must adhere to certain formats and standards. Since a primary application of SHM platforms is in real-time data analysis, the platforms need to be able to process the data gathered by various sensors for which the data should be in compatible formats for the platform to convert into usable information.
- **System Compatibility:** When integrating the sensor hardware and SHM software, it is important to ensure they are compatible with each other. This can range from stepwise harmonizing of the global structures of systems and their interfaces to allow proper interaction of various elements of a system.
- **Real-Time Data Processing:** To aid monitoring, real-time data processing capabilities must be incorporated into the SHM system. This makes it possible to perform real-time analysis and, in turn, respond to the results obtained from IoT sensors.

4.6 ADVANCED DATA ANALYTICS FOR STRUCTURAL MONITORING

Improving the monitoring of structures requires a return to data and analytics, taking the information that we receive from sensors and turning it into something useful. This means that the measures are more accurate and can predict maintenance work on an organization's infrastructure better through ML, AI technology for predictive maintenance, and cloud-based platforms. So, let us thoroughly explore about these advanced data analytics, and how they have changed the dynamic of SHM (Cao et al., 2020).

4.6.1 Machine Learning and Its Uses in Artificial Intelligence for Data Analysis

These are two great techniques, called AI and ML, that can not only modify the way analysis has been done formerly but also help systems to learn from data error and perform better than before. Algorithms of ML and AI are applied to a large quantity of information made available from product monitoring by several IoT devices in structural monitoring and learn how to detect patterns, faults, and trends which can be analysed plainly.

- **Anomaly Detection:** Depending on the analysis of historical data, an ML algorithm can recognize signs of abnormalities and problems. For instance, AI models can be trained to detect abnormal vibration, both internal and external and stress levels that may point towards structural faults.
- **Pattern Recognition:** AI can work on large sets of data and determine such patterns among the data that may be repeated and have correlations. This

is also useful in comprehending how various rates are associated with one another, for example, the effects of temperature variation on the stress within the structure.

- **Decision Support:** There is a claim that the use of intelligent systems can allow for making recommendations stemming from the analysis of the available data. For example, AI can propose preventive or redesign actions to be made in response to signs that arise and to problems that may become critical.

4.6.2 Predictive Maintenance Algorithm

Suddenly, I discovered that algorithms for predictive maintenance employ data analysis to determine when maintenance activities have to be carried out before failure in order to reduce the lifecycle of infrastructure. These algorithms use prior experience and data acquired from vehicle sensors to forecast conditions and requirements for impending service.

- **Failure Prediction:** Predictive models use data to identify trends in order to determine possible failure points. For instance, there is a case of stress history and load conditions under which machine algorithms can estimate when a bridge will likely need repairs.
- **Maintenance Scheduling:** These algorithms schedule preventive maintenance activities by anticipating when components are most likely to fail. This approach greatly reduces the frequency of outages and the time they last.
- **Resource Optimization:** Predictive maintenance is especially useful in improving management of the scarce resources available since maintenance is carried out only when necessary. This in turn reduces cost related to inspections and repairs of structures which are not really required.

4.6.3 Cloud-Based Platforms for Remote Monitoring and Analysis

- Advanced cloud solutions are becoming an element of modern SHM systems because cloud solutions provide enough potential and adaptability for data storage and analysis. These environments allow one to access data from a distance and make decisions with other stakeholders.
- **Data Storage and Management:** Cloud networks offer tremendous storing capacities for the huge amount of data produced by the IoT sensors. Such central data storage enables efficient storage of historical and real-time data, hence easing management.
- **Remote Access and Analysis:** One of the most significant features of cloud-based systems is the possibility of engineers and analysts to analyse

data wherever they can access the internet. This flexibility is especially valued in cases of large construction projects or in situations when it is necessary to control an area located at a great distance.

- **Scalability and Integration:** Cloud provides versatile solutions that allow the accommodation of large volumes of data and interaction with different analytics and technologies. Such prototyping guarantees the practical applicability of the monitoring system in light of growing stream and analytical loads.
- **Collaboration and Sharing:** Web-based systems enable real-time working they provide access to most stakeholders and allow working on data in more than one place at a time. This improves decision-making, and coordination between working teams is effective. Hence, all the stakeholders get the best information.

4.6.4 Wireless Communication Protocols in IoT Structural Monitoring

Organizing SHM systems applying IoT is reliant on wireless communication standards. They facilitate the transfer of the collected data through the sensors to the monitoring platforms and reliably report any valuable information relating to the health status of any infrastructures. Within this section, the utilization of different types of wireless technologies, energy efficiency, data transfer rates, and main parameters of network security and availability are discussed (Safa et al., 2016).

4.6.5 Role of Wireless Technologies (LoRa Zigbee 5G)

Wireless technologies are central to the operation of IoT systems, each offering unique advantages for different monitoring scenarios:

- **LoRa (Long Range):** LoRa is now famous for its long-range communication and low power consumption. It is primordially applicable where there are several sensors spread over large sections as is the case in monitoring of bridges or dams. Due to features such as transmitting data over several kilometres, LoRa is suitable for remote and low maintenance applications. But it sustains lower data rates than other technologies, and applications which involve high frequency rate of data update are not feasible.
- **Zigbee:** Zigbee is intended for applications based on the restricted transmission range and works at the frequency of 2.4 GHz. It is ideal for applications that require low power levels combined with low data rates forming SHM of buildings. Zigbee network is normally established in a mesh topology which gives it versatility and reliability through layering of data throughout

the network. This makes Zigbee ideal for those places where the proximity between the sensor nodes is fairly reasonable.

- **5G:** The new generation of wireless systems, 5G, is designed to achieve high data rates and low latencies, which is suitable for big data system applications. In structural monitoring, 5G can enable high-definition video surveillance and extensive data gathering from many sensors. It is advantageous to it [the use of social media for business intelligence] in that its bandwidth is wider and its latency is lower in contexts requiring instant response and minute data.
- **Energy Efficiency:** The IoT sensors that can be positioned in areas that are distant must be energy efficient. Both LoRa and Zigbee technologies are low-power technologies, meaning sensors and other connected devices can run for long on battery power. Low-energy communication protocols ensure that the network does not require constant maintenance and battery replacements, hence saving cost.
- **Data Transmission Rates:** The actual rate of data transmission depends on the kind of data to be transmitted and the application to be made. For instance, high-frequency data from sensor or high-definition images requires higher data rates, which technologies like 5G offer. On the other hand, it is possible to achieve the opposite objective—use LoRa or Zigbee with lower transmission rates in applications requiring less amounts of data. In order to reach efficient system operation, it is possible to not only set a high data transfer rate but also to increase the energy efficiency of the communication technology.

4.6.6 Security and Privacy in IoT-Based Structural Monitoring

The security and privacy of the obtained data are crucial when IoT plays an increasingl significant role in structural monitoring. The ability to counter various security threat forms part of the effectiveness of these systems as much as the technological aspect o the systems. As part of this review, the privacy and security aspects fundamental t IoT-based structure monitoring systems are presented.

Safeguarding Information Integrity and Avoiding Sensor Manipulation Dat Integrity: It is crucial to pay close attention to the quality and reliability of data collected from IoT sensors. Biased or inaccurate information may lead to incorrect structural health determinations, which may in turn lead to dangerous consequences o inadequate measures. When it comes to editorial issues, it is, therefore, mandatory t prevent any manipulation and provide very thorough error-checking techniques.

Preventing Sensor Tampering: Obviously, physical intervention influence how effectively sensors operate and the quality of the information collected. This risk requires the use of a safe zone and limited access to the area where sensors are installed Other preventions of any attempt to tamper with the system include incorporating tampering indicators and contracting maintenance inspection services.

Authentication and Access Control: Access to the sensible data stored in sensors and the necessary configurations must be protected, which is why setting up the proper type of identification mechanisms is inevitable. Multi-factor authentication (MFA) and role-based access control (RBAC) are worthwhile strategies for preventing unauthorized entrance and unauthorized tampering of sensor system.

Secure communication in IoT networks and protection from unauthorized access and wiretapping need to have encryption of the data exchange between IoT sensors and the central systems. To ensure the security of data at the time of transmission and to ensure that it is not tempered/modified, TLS (transport layer security) should be used.

Utilizing firewalls and IDs as effective methods of running secure networks is fundamental for preventing attacks on IoT networks. Operating system and software updates and patch installation on software and network devices frequently safeguard against risks that attackers may take control of.

Secure Data Storage: In addition to protecting data during its transfer, stored data located on databases or in cloud services must also be protected. Risks that can be eliminated through data backups include the following: Data backup assures the security and reliability of recovery procedures.

Case Study: Urban Secure IoT Monitoring for Downtown Development: A System Synopsis: The integration of SHM for a city's downtown is another great and secure IoT monitoring technique. This project was about trying to maintain reliability and safety in critical urban infrastructures such as tunnels and high-rise buildings and bridges.

Implementation: Secure Sensor Deployment: Human sensors were installed in secure boxes, while physical sensors were attached securely to a variety of urban structures which were lockable. To avoid such incidences, we did normal maintenance checks to identify any flaws or interference.

Safeguards, in terms of data security, were considered and effected such as encryption of the data that these sensors gathered and archived. With the push for more secure and encrypted measures for data communication, protocols such as TLS and other secure communications protocols saved data from illegal access and eavesdropping.

As a result of the strict access rights set up by RBAC as well as the introduction of the MFA, it was possible to develop a centralized monitoring system. This system allowed the concerned or authorized people to get direct access to the sensor data with trail of audit for accessing and changing.

4.7 NETWORK SECURITY AND RELIABILITY

Ensuring network security and reliability is paramount in IoT-based structural monitoring systems, as they handle critical data related to infrastructure health:

Network Security: Efforts must be made to preserve data's unity and secrecy in order to maintain the data from being controlled and modified. Measures which have been put in place include protection of data transmitted in networks through an encryption process, provision of secure means of authentication for the devices to be used, and the users, as well as periodic updates on security loopholes for their fixing. Additionally, these measures encompass continuous monitoring for potential threats, implementation of access control mechanisms, adherence to cybersecurity best practices, and regular security audits to ensure compliance with evolving standards and regulations.

Reliability: The accomplishment of these goals depends heavily on the availability and reliability of the communication network for data transmission. Redundancy, error correction, and network monitoring are some of the methods that can be used to improve reliability. For instance, Zigbee's mesh connection makes it easier to have multiple paths through which information may be conveyed, and 5G's network framework fosters high availability and network redundancy.

4.8 EDGE COMPUTING IN IoT-ENABLED STRUCTURAL MONITORING

Remote, or edge, computing has become one of the most significant innovations for IoT applications in structural monitoring, offering impressive gains in performance, lower latency, and better organizational organization. With the proffered aspects of computation and data analysis at the site of data originating, edge computing holds a number of advantages in handling and evaluating structural health. This chapter provides further information on what are edge devices, how latency can be minimized, and what edge computing means for smart buildings (Stojanović et al., 2021).

4.8.1 The Use of Edge Devices in On-Site Data Analytics

Local computing involves computation and analysis to be done at the site where data is produced, and edge devices are key to doing this. This decentralization of processing power offers several advantages:

- **Local Data Processing:** Collectively, edge devices can analyse and manipulate data collected from the sensors without the need to transfer voluminous data to centralized servers or the cloud for processing. This probably facilitates the almost real-time decision-making, especially concerning structural conditions since the processing is localized.
- **Reduced Bandwidth Usage:** Edge devices allow the actual data processing to be centralized because they reduce the amount of data that needs to be transmitted over the network. It is hoped that such decreased information

flow will lessen the load within the network, leading to lesser operational expenses in terms of data transport.

- **Enhanced Data Privacy:** Local processing helps keep all data on-site, which increases its security and confidentiality. Most importantly, a monitoring system will be useful where sensitive data is not available in the public domain, such as trade secrets.
- **Operational Efficiency:** First, distributed edge devices can pre-process some basic data extraction and summarization that help reduce the amount of data that flows to central systems. This enhances the operation of data storage and analysis methods.

4.9 REDUCING LATENCY IN STRUCTURAL HEALTH MONITORING

These losses represent an essential concept of SHM, known as latency, or the time it takes for the data to be created and analysed. Reducing latency is essential for timely responses to potential structural issues:

- **Real-Time Analysis:** The general concept of edge computing is the computation of real-time data processing at or as close as possible to the source of information. This saves time for data to reach a control server and then back to identify structural irregularities and respond to them.
- **Immediate Feedback:** When collecting information on the state of the structure, edge devices avoid feedback on the state of the structure due to the transfer of data to other sites or devices. This is important for operations like dynamic load testing or measurement when conditions like storming are raging and quick outcomes are paramount to safety.
- **Optimized Decision-Making:** Low latency is valuable because it helps in decision-making since the data is current. Maintenance planners can provide engineers and maintenance teams with the latest data to reduce possible problems and enhance maintenance planning and programs.
- **Deployment of Edge Devices:** Accompanying sensors such as accelerometers, strain gauges, and thermometers were fitted all around the building to enable edge devices to capture relevant data from them. These devices were installed in strategic areas of beams, columns, and foundations of the structural system.
- **Local Data Processing:** The first two tasks were processed and analysed at the edge devices. For example, they blocked ambiguous signals from acquired sensors and combined data from several sensors which gave proper results regarding the health of the building.

- **Real-Time Monitoring and Alerts:** By instantly sharing information on structural problems, the edge devices allowed for real-time monitoring. The edge devices immediately alerted maintenance workers to any abnormalities discovered by the sensors, such as unexpected vibrations or strain levels, enabling prompt action if needed.
- **Improved Operational Efficiency:** The system optimized network use and storage by processing data locally, which decreased the amount of data transported to the central server. This made it easier to analyse long-term data and historical patterns using cloud resources more effectively.
- **Better Building Management:** By integrating edge computing, the management system of the smart building was able to use real-time data to make better-informed choices. This included making modifications to building operations, depending on the existing structural issues, such as changing the heating, ventilation, and air conditioning system or redistributing the load.

4.10 CHALLENGES IN IoT SENSOR DEPLOYMENT FOR STRUCTURAL HEALTH MONITORING

Monitoring structural health using IoT sensors brings about some issues, which the structural engineers must clearly understand. Thus, awareness of these issues and the possibility of realizing relevant measures can significantly improve the efficiency of structural monitoring. We now discuss the main issues of sensor calibration and drift, power management and battery, as well as influences of the environment and weather conditions on the sensors (Rossi et al., 2016).

4.11 ENVIRONMENTAL AND WEATHER IMPACTS ON SENSOR PERFORMANCE

Weather-Related Issues: Environmental conditions such as temperature, humidity, and precipitation can affect sensor functionalism. Sensors that face severe environmental conditions could have problems from condensation, corrosion, and thermal expansion, leading to reduced accuracy and reliability.

Weatherproofing/IP Rated Components: In order to protect them from the environment, sensors will be housed in appropriate enclosures. These enclosures should be able to withstand outdoor conditions and allow for adequate ventilation so they do not overheat or build up condensation.

Environmental calibration sensors must be calibrated to the prevailing environment. For example, a temperature-compensated sensor will compensate for changes in ambient temperature by adjusting their reading and giving better measurements.

Low long-term durability sensors that can last in these environments for years is a critical part of any IoT mission, which silicon-based devices typically cannot do. This means: the choice of durable, weatherproof, and solid materials and forms to ensure structural integrity, longevity, and resilience against environmental factors.

- **Comprehensive Testing:** Engaging in exhaustive sensor testing across diverse environmental conditions is crucial for pinpointing potential issues and enhancing design specifications. This rigorous testing guarantees consistent sensor performance across the anticipated array of conditions.
- **Adaptive Maintenance Strategies:** Deploying adaptive maintenance approaches, like remote monitoring and diagnostics, aids in the early identification of calibration, power, and environmental issues. This pre-emptive strategy enables prompt interventions and maintenance procedures.
- **Integration with Monitoring Systems:** Integrating sensors with advanced monitoring systems that can detect and alert for potential calibration drifts or power issues enhances overall system reliability. These systems can provide real-time data on sensor status and performance, facilitating prompt corrective actions.

4.12 NEW TRENDS IN SENSOR SYSTEMS

- **Nano-Sensors:** Thus, the nano-sensors significantly improve the sensor technology owing to their high sensitivity and compact size. These sensors also precisely detect small changes in structural features such as temperature, stress, or strain, among other factors due to nanotechnology. Nano-sensors could be introduced into materials simply because they have a near molecular size that can point out all the structural defects. Their use may lead to methods of monitoring that are less obtrusive, yet more effective.
- **Smart Materials:** Interestingly, another praiseworthy aspect of this approach is its ability to enhance efficiency and accuracy in complex problem-solving scenarios. These materials could enhance structural durability, improve sustainability, and optimize construction efficiency in various engineering applications. For example, piezoelectric materials generate electrical charges when subject to mechanical stress and thus allow for real-time measurement of load and deformation on structures. The opportunity for the dynamic integration of structural technologies for monitoring and structural health has its basis in the use of IoT sensors and SMs.

4.13 CONCLUSION

To sum up, the introduction of IoT sensors in the structural status of the building is
revolution in the solution of geotechnical engineering. In this chapter, we have consi
ered the importance of its role in SHM and management to extend the lifespan as well a
safety of structures, which is the core of IoT technology and networks. In that capacit
various IoT sensors may be utilized by engineers to fill an ideal toolbox for accurate an
timely data gathering. These sensors consist of deformations gauges, load estimatio
sensors such as pressure sensors and vibration, and tilt sensors such as accelerometers

These sensors provide useful performance-related attributes, existing inside struc
tures and configured in a centralized or distributed mode, regularly monitoring th
critical locations, multimedia window. Data integration frameworks and data collectio
systems contain complex algorithms and mechanisms which make these technologie
substantially more efficient for predictive analytics and data integration. Thus, this syn
ergy is crucial for arranging prompt treatments and actualized prevention of failures.

The practical overlays of IoT sensors are clearly illustrated through extensive exe
cises, including catastrophe response systems and bridge health monitoring. Othe
advanced innovations, including digital twins and nano-sensors, are expected t
enhance structural monitoring in the future. IoT will without doubt offer novel opportu
nities to enhance structure integrity and vulnerability, as it progresses; therefore, it wi
shape geotechnical engineering.

REFERENCES

Cao, Y., Miraba, S., Rafiei, S., Ghabussi, A., Bokaei, F., Baharom, S., & Assilzadeh, H. (2020
Economic application of structural health monitoring and internet of things in efficiency o
building information modeling. *Smart Structures and Systems, 26*(5), 559–573.

Dackermann, U., Crews, K., Kasal, B., Li, J., Riggio, M., Rinn, F., & Tannert, T. (2014). In sit
assessment of structural timber using stress-wave measurements. *Materials and Structures*
47, 787–803.

Hajovsky, R., & Pies, M. (2012). Measuring system for longtime monitoring and visualization o
temperature and toxic gases concentration. *Elektronika ir Elektrotechnika, 122*(6), 129–132

Hassani, S., Mousavi, M., & Gandomi, A. H. (2021). Structural health monitoring in composit
structures: A comprehensive review. *Sensors, 22*(1), 153.

Hérail, P., & Bit-Monnot, A. (2023). Leveraging demonstrations for learning the structur
and parameters of hierarchical task networks. *The International FLAIRS Conferenc*
Proceedings, 36, (pp. 1–7).

Ma, C., Huang, J. B., Yang, X., & Yang, M. H. (2015). Hierarchical convolutional features fo
visual tracking. In *Proceedings of the IEEE International Conference on Computer Visio*
(pp. 3074–3082). Santiago, Chile

Mehrabi, P., Honarbari, S., Rafiei, S., Jahandari, S., & Alizadeh Bidgoli, M. (2021). Seismic response prediction of FRC rectangular columns using intelligent fuzzy-based hybrid metaheuristic techniques. *Journal of Ambient Intelligence and Humanized Computing, 12*, 10105–10123.

Moreu, F., Jo, H., Li, J., Kim, R. E., Cho, S., Kimmle, A.,... LaFave, J. M. (2015). Dynamic assessment of timber railroad bridges using displacements. *Journal of Bridge Engineering, 20*(10), 04014114.

Rossi, M., Calderini, C., & Lagomarsino, S. (2016). Experimental testing of the seismic in-plane displacement capacity of masonry cross vaults through a scale model. *Bulletin of Earthquake Engineering, 14*, 261–281.

Sadhu, A., Peplinski, J. E., Mohammadkhorasani, A., & Moreu, F. (2023). A review of data management and visualization techniques for structural health monitoring using BIM and virtual or augmented reality. *Journal of Structural Engineering, 149*(1), 03122006.

Safa, M., Shariati, M., Ibrahim, Z., Toghroli, A., Baharom, S. B., Nor, N. M., & Petković, D. (2016). Potential of adaptive neuro fuzzy inference system for evaluating the factors affecting steel-concrete composite beam's shear strength. *Steel and Composite Structures, An International Journal, 21*(3), 679–688.

Singh, P., Mittal, S., & Sadhu, A. (2023). Recent advancements and future trends in indirect bridge health monitoring. *Practice Periodical on Structural Design and Construction, 28*(1), 03122008.

Stojanović, J., Petkovic, D., Alarifi, I. M., Cao, Y., Denic, N., Ilic, J., & Milickovic, M. (2021). Application of distance learning in mathematics through adaptive neuro-fuzzy learning method. *Computers & Electrical Engineering, 93*, 107270.

IoT for Conservation Requirement Examining the Foundation of Households

5

5.1 INTRODUCTION TO CONSERVATION REQUIREMENTS IN HOUSEHOLDS

This chapter presents a major revolution in the aspect of Internet of Things (IoT) integration as part of efforts in the conservation of households. In our effort towards sustainability, it has dawned on human beings that resource conservation is paramount, hence IoT provides distinct solutions to a range of basic necessities such as water, energy, and waste. The utilization of smart water metres and the integration of leakage detection systems demonstrate how the use of real-time monitoring techniques can help minimize the wastage of water while the same degree use of smart metres and energy-efficient home electrical appliances shows how IoT can be used to minimize electricity usage (Alcubilla & Lund, 2006).

DOI: 10.1201/9781003537625-5

5.1.1 The Role of IoT in Enhancing Household Efficiency

This presents a major revolution in the aspect of IoT integration as part of efforts in the conservation of households. In our effort towards sustainability, it has dawned on human beings that resource conservation is paramount hence IoT provides distinct solutions to a range of basic necessities such as water, energy, and waste. The utilization of smart water metres and integration of leakage detection systems demonstrates how use of real-time monitoring techniques can help minimize the wastage of water while to the same degree use of smart metres and energy-efficient home electrical appliances shows how IoT can be used to minimize electricity usage (Walie & Fisseha, 2016).

The role of IoT is not just monitoring; IoT optimizes household performance through smart sensors, smart actuators, and smart connectivity. Advanced technology offers more concrete recommendations on how resources are being used, thereby allowing the homeowners to act and conserve on their own. Real-life stories from urban households and smart homes demonstrate that such technologies give a real positive shift in terms of resource efficiency and sustainability.

5.1.2 Resource Conservation in Households

The role of IoT is not just monitoring; IoT optimizes household performance through smart sensors, smart actuators, and smart connectivity. Advanced technology offers more concrete recommendations on how resources are being used, thereby allowing homeowners to act and conserve on their own. Real-life stories from urban households and smart homes demonstrate that such technologies give a real positive shift in terms of resource efficiency and sustainability.

But the process of achieving IoT-assisted conservation paradigms is not without obstacles. Factors such as technical constraints, funding shortfalls, and deficiencies in the built environment as well as data confidentiality issues and system compatibility questions likewise remain a concern so that it is crucial to increase the learnings to populations in order to be most fruitful. In the future, the trends including the connection of IoT with renewable energy, edge computation, and artificial intelligence (AI) will unlock new innovations in household conservation (Xiao et al., 2018).

Lastly, IoT becomes a solution to support sustainable living since the households can control the resources they use intelligently. Exploiting the theoretical framework that can be derived from the present case, it is possible to open the further evolution and development of IoT with correct accomplishments laying the foundation for a greener and more sustainable future.

5.1.3 Importance of Conservation for Sustainable Development

But the process of achieving IoT-assisted conservation paradigms is not without obstacles. Factors such as technical constraints, funding shortfalls, and deficiencies in the built environment as well as data confidentiality issues and system compatibility questions likewise remain a concern so that it is crucial to increase the learnings to populations in order to be most fruitful. In the future, the trends including the connection of IoT with renewable energy, edge computation, and AI will unlock new innovations in household conservation.

Lastly, IoT becomes a solution to support sustainable living since the households can control the resources they use intelligently. Exploiting the theoretical framework that can be derived from the present case, it is possible to open the further evolution and development of IoT with correct accomplishments laying the foundation for a greener and more sustainable future (Fisseha et al., 2011).

5.1.4 Fundamentals of IoT in Household Resource Management

IoT is simply a system that comprises material objects that can interact with each other via the Internet. These, at times sensor and actuator-integrated devices collect, send, and even respond to data for a given task to be self-executed or with very little human interaction required. In the framework of the IoT applications for resource management in homes, objects, and household devices join a networked environment and become smart, which helps to optimize the use of potentially limited resources including water, energy, and waste (Table 5.1).

5.2 SENSORS, ACTUATORS, AND NETWORKING IN HOME AUTOMATION IoT NETWORKS

Resources in household IoT systems are closely managed by the sensors and actuators of the system. Measuring devices that track several attributes connected with resource utilization are called sensors. For instance, smart water metres utilize smart flow sensors to monitor and quantify water utilization, while smart temperature sensors in smart thermostats observe and control the setting's climate. These sensors offer current information about how homeowners utilize the resources within their homes (May et al., 1998).

TABLE 5.1 Fundamental of household resource management

FIELD	*COMPONENT*	*DESCRIPTION*	*ADVANTAGES*	*DISADVANTAGES*	*USE CASES*
Energy management	Smart metres	Smart metres monitor electricity usage in real-time, enabling users to track consumption and optimize energy usage based on data-driven insights	Provides real-time feedback, helps in reducing energy costs, and supports energy efficiency efforts	Potential privacy concerns as smart metres track detailed household energy consumption patterns	Residential energy management, peak-demand control, energy-saving programs in smart homes
Water conservation	IoT-enabled water sensors	Water sensors monitor flow, detect leaks, and optimize water usage in household appliances such as washing machines, showers, and sprinklers	Reduces water wastage, provides early detection of leaks, and supports efficient irrigation in smart gardens	May require regular maintenance to ensure accuracy and can be costly to install initially	Leak detection systems, automated irrigation in gardens, smart water-saving shower systems
Appliance control	Smart plugs and smart appliances	Smart plugs and appliances allow users to control household devices remotely, schedule operations, and optimize energy consumption through IoT-enabled connectivity and automation	Allows remote access, energy-efficient scheduling, and automation of household appliances, resulting in energy savings	Security risks such as hacking can compromise control of appliances, and reliance on stable Internet connections is necessary for smooth operation	Smart kitchen appliances (e.g., refrigerators, ovens), remotely controlled lighting systems, automated home heating and cooling

(Continued)

TABLE 5.1 (*Continued*) Fundamental of household resource management

FIELD	*COMPONENT*	*DESCRIPTION*	*ADVANTAGES*	*DISADVANTAGES*	*USE CASES*
Waste management	IoT-based waste bins	IoT-enabled smart bins track the fill level of waste containers and optimize waste collection schedules for efficient resource management in households	Improves waste management efficiency, reduces waste collection costs, and supports environmental sustainability by encouraging responsible disposal	Initial cost of IoT bins may be high, and integration with municipal waste systems may require additional infrastructure	Smart waste bins that notify when full, waste management systems in smart cities, and composting tracking in residential areas
Security and surveillance	Smart security systems (CCTV, door sensors)	IoT-enabled security systems monitor household activity, provide real-time alerts, and enable remote control over home security features such as door locks, cameras, and alarms	Enhances home security with real-time notifications, remote access, and integration with smart home ecosystems	Security breaches or hacking risks if not adequately protected, potential loss of privacy if surveillance data is not securely managed	Smart doorbells with cameras, motion detection security cameras, smart locks, and alarm systems integrated with IoT platforms

Conversely, actuators make active responses to the data obtained by the sensors by performing tasks as commanded. For example, Smart irrigation system can employ actuators to open/close water valves or even regulate watering timing according to the moist level arising from the moisture sensors. The direct talk of the sensors with the actuators makes it easier for a home to be automated and managed in an efficient manner as far as the resources it possesses are concerned.

5.3 WATER CONSERVATION USING IoT SYSTEMS

IoT is simply a system that comprises material objects that can interact with each other via the Internet. These, at times sensor and actuator-integrated devices collect, send, and even respond to data for a given task to be self-executed or with very little human interaction required. In the framework of the IoT applications for resource management in homes, objects, and household devices join a networked environment and become smart, which helps to optimize the use of potentially limited resources including water, energy, and waste

$$\text{Objective}: W(t) = \sum_{i=1}^{N}\left(\alpha_i \cdot \frac{C_i(t)}{R_i(t)} + \beta_i \cdot \frac{D_i(t)}{S_i(t)}\right)$$

$$C_i(t) = \int_0^T \frac{F_i(t)}{\log(1 + L_i(t))} \cdot \frac{\partial^2 P_i(t)}{\partial t^2}\, dt$$

$$R_i(t) = \frac{\int_0^T Q_i(t) \cdot \frac{\partial F_i(t)}{\partial t}\, dt}{1 + \log(1 + L_i(t))}$$

$$D_i(t) = \frac{\partial^2 Q_i(t)}{\partial t^2} + \lambda \cdot \frac{\partial^3 Q_i(t)}{\partial t^3}$$

$$S_i(t) = \frac{\int_0^T F_i(t) \cdot \frac{\partial P_i(t)}{\partial t}\, dt}{\log\left(1 + \frac{\partial P_i(t)}{\partial t}\right)}$$

$$L_i(t) = \sum_{j=1}^{M}\left(\frac{\partial^2 P_{ij}(t)}{\partial t^2} + \mu \cdot \frac{\partial^3 P_{ij}(t)}{\partial t^3}\right)$$

$$P_i(t) = \frac{\partial}{\partial t}\left(\frac{\partial F_i(t)}{\partial t} + \lambda \cdot \frac{\partial^2 F_i(t)}{\partial t^2}\right)$$

$$Q_i(t) = \frac{1}{\log(1+F_i(t))} \cdot \frac{\partial F_i(t)}{\partial t}$$

$$F_i(t) = \sum_{k=1}^{N} \left(\frac{\partial}{\partial t} (W_k(t) \cdot \theta_k(t)) \right)$$

$$\frac{dW(t)}{dt} = \sum_{i=1}^{N} \left(\alpha_i \cdot \frac{dC_i(t)}{dt} + \beta_i \cdot \frac{dD_i(t)}{dt} \right)$$

$$\frac{dC_i(t)}{dt} = \frac{\partial}{\partial t} \left(\int_0^T \frac{F_i(t)}{\log(1+L_i(t))} \cdot \frac{\partial^2 P_i(t)}{\partial t^2} dt \right)$$

$$\frac{dR_i(t)}{dt} = \frac{d}{dt} \left(\frac{\int_0^T Q_i(t) \cdot \frac{\partial F_i(t)}{\partial t} dt}{1 + \log(1+L_i(t))} \right)$$

$$\frac{dD_i(t)}{dt} = \frac{d}{dt} \left(\frac{\partial^2 Q_i(t)}{\partial t^2} + \lambda \cdot \frac{\partial^3 Q_i(t)}{\partial t^3} \right)$$

$$\frac{dS_i(t)}{dt} = \frac{d}{dt} \left(\frac{\int_0^T F_i(t) \cdot \frac{\partial P_i(t)}{\partial t} dt}{\log\left(1 + \frac{\partial P_i(t)}{\partial t}\right)} \right)$$

Symbols Explanation:

- $W(t)$ is the total water conservation metric at time t.
- $C_i(t)$ is the water consumption rate of the system i at time t.
- $R_i(t)$ is the recovery rate or recycled water amount for system i.
- $D_i(t)$ is the drainage or loss of water from system i.
- $S_i(t)$ is the system sustainability measure, balancing water use and savings.
- $F_i(t)$ is the flow rate of water in system i at time t.
- $P_i(t)$ is the pressure rate of the system i at time t.
- $Q_i(t)$ is the water quality index for system i.
- $L_i(t)$ is the leakage rate in the system i at time t.
- α_i, β_i is the weighting factors for different components of water management.
- λ, μ is the regularization parameters.
- N is the number of IoT systems in the network.
- M is the number of water pressure or leakage measurement points in system i.

5.3.1 Real-Time Alerts and Water Usage Insights

IoT is simply a system that comprises material objects that can interact with each other via the Internet. These, at times sensor and actuator-integrated devices collect, send, and even respond to data for a given task to be self-executed or with very little human interaction required. In the framework of the IoT applications for resource management in homes, objects, and household devices join a networked environment and become smart, which helps to optimize the use of potentially limited resources including water, energy, and waste (Powrie & Beaven, 1999).

5.4 ENERGY EFFICIENCY AND POWER CONSUMPTION MONITORING

A prime example is the smart metres which are veering the consumption of electrical power in the households by offering fine, current information on the power usage. Smart metres differ from ordinary metres that only provide the customer with occasional readings: the function of smart metres is to constantly measure the electricity usage and share the collected data not only with the utility company but also with the homeowner. This makes it possible for electricity billing to be done following the intensity of power used in place of guesswork and it enables users to have measurable energy intensity (Figure 5.1).

5.4.1 Smart Metres of Electricity Consumption

Thus, the liberal use of smart metres goes beyond the efficiency of the billing procedures. They give information on consumers and guide households in the identification of peak demands as well as power-hungry devices. This analysis will enable homeowners to understand what particular aspect of the house uses the most electricity, and the necessary adaptations that can be made. For example, if a smart metre shows that in certain hours of the day you use more energy, then everybody in the household can change their habits, or plan to do activities that require energy during a time of day when the energy is cheaper (Alemu et al., 2011).

Smart IoT-connected home appliances are built to improve the consumer's efficiency of the device to save energy. These include refrigerators, washing machines, and thermostats, among others, these appliances have sensors and connectivity tools that enable them to work depending on the data collected. For instance, a smart thermostat helps to understand the temperature habits of an apartment and regulate settings to save energy during everyone's absence.

Besides automation, IoT home appliances are capable to connect to other smart devices in the home to generate energy-saving solutions. For example, a washing machine may be incorporated with a facility that allows it to wash clothes during times

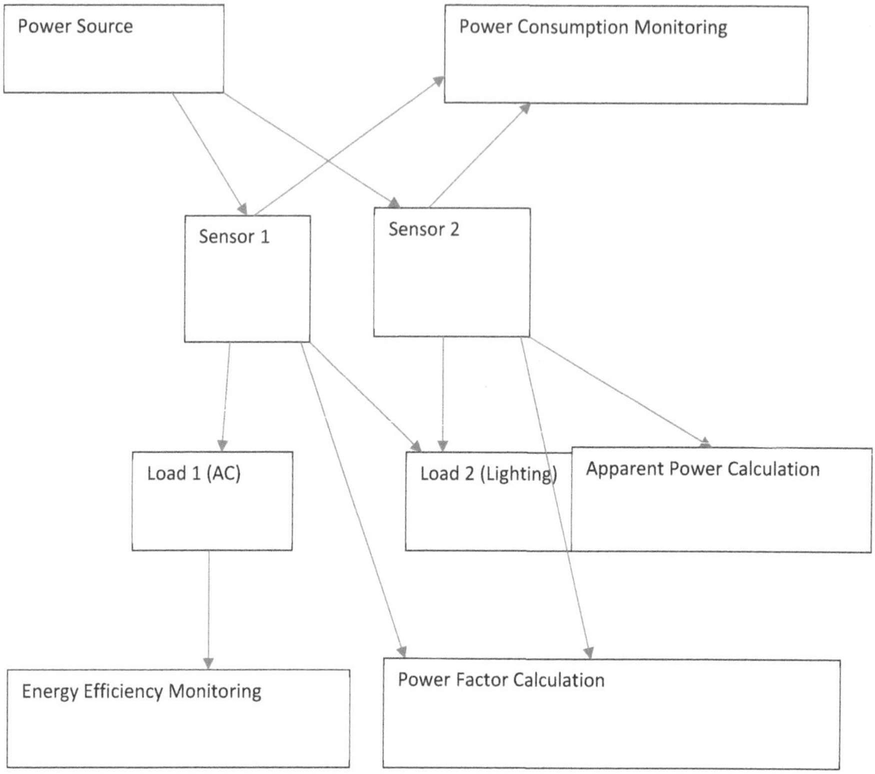

FIGURE 5.1 Energy efficiency and power consumption monitoring.

of low electricity usage or a refrigerator may contain an ability to control the appropriate cooling cycles depending on the regularity of usage. In this level of integration, not only is the energy consumption cut down, but also convenience and experience of the end user is improved as well.

5.4.2 Optimizing Energy Use through Real-Time Monitoring

A prime example is the smart metres which are veering the consumption of electrical power in the households by offering fine, current information on the power usage. Smart metres differ from ordinary metres that only provide the customer with occasional readings: the function of smart metres is to constantly measure the electricity usage and share the collected data not only with the utility company, but also with the homeowner. This makes it possible for electricity billing to be done following the intensity of power used in place of guess work and it enables users to have measurable energy intensity (Boafo et al., 2016).

5.4.3 Smart Metres of Electricity Consumption

Thus, the liberal use of the smart metres goes beyond the efficiency of the billing procedures. They give information on consumers and guide households in the identification of peak demands as well as power-hungry devices. This analysis will enable homeowners understand what particular aspect of the house uses most electricity, and the necessary adaptations can be made. For example, if a smart metre shows that in certain hours of the day you use more energy, then everybody in the household can change their habits, or plan to do activities that require energy during a time of day where the energy is cheaper.

Smart IoT-connected home appliances are built for improving the consumer's efficiency of the device to save energy. These include refrigerators, washing machines, thermostats among others, these appliances have sensors and connectivity tools that enable them work depending on the data collected. For instance, a smart thermostat helps to understand the temperature habits of an apartment and regulate settings to save energy during everyone's absence.

Besides automation, IoT home appliances are capable to connect to other smart devices in the home to generate energy-saving solutions. For example, a washing machine may be incorporated with a facility that allows it to wash clothes during times of low electricity usage or a refrigerator may contain an ability to control the appropriate cooling cycles depending with the regularity of usage. In this level of integration, not only is the energy consumption cut down, but also convenience and experience of the end user is improved as well.

5.5 WASTE MANAGEMENT WITH IoT SOLUTIONS

Smart waste bins are not only meant for sorting waste but also have a feature that enables them to report issues to do with the fill levels to waste collection agencies. When a bin or bin location is about to be full, the waste collection service is alerted regarding bin location and route optimization. Such real-time monitoring cuts the number of random pickups and assures that bins are emptied right on time hence enhancing effectiveness and cutting cost of operation in waste collection.

5.5.1 IoT for Smart Waste Segregation and Collection Systems

Smart waste segregation and collection have been discovered by the integration of IoT in waste management. Advanced waste baskets that are fitted with sensors and connectivity interfaces can sort the waste in bins into various classes including recy clable waste, compostable waste, and general waste. There are bins here that emplo sensors that identify the kind of waste that is being disposed of as well as appropria compartmentalization. But it also helps to mitigate contamination of recycling stream which can enhance the quality of recycled materials.

5.5.2 Minimizing Waste through IoT-Based Activity Monitoring Deutsch

The consumption patterns, waste generation, and overall smart conservation throug monitoring based on IoT systems contribute a great deal to low wastage in a househol Many smart kitchen appliances like smart refrigerator and smart food waste monito can be useful for proper handling of foods in a home. For example, it can be sma refrigerator that checks the availability and expiration date of foods at home and alert that it is time to use them. This assist in reducing cases of food wastage and helps i meal optimization.

IoT sensors can also be placed on wastepaper baskets in order to identify the kind of waste produced and the volume of waste manufactured. The collected data can the be used to find recurrent issues so that optimization efforts can be directed appropri ately. For instance, if a household that regularly disposes off many items of food, it wil lead to modification of shopping pattern or cooking habits (Sønderlund et al., 2016).

5.5.3 Applications of AI and IoT in Household Waste Reduction

The blend of AI and IoT technologies provides great home waste management applica tions. They involved leveraging AI algorithms to sift through the data collected from IoT devices to understand potential trends, and making recommendations related t waste management improvements. This means, for instance, machine learning model predicting future waste collections by further harnessing previous waste productio trends.

In addition, AI-powered systems can help to enhance the functionality of smar waste bins in terms of waste-sorting accuracy. To help identify and sort various type of waste more easily, image recognition and machine learning can be enabled in such systems to achieve optimum recycling operations. AI-driven analytics can also supply tailor-made suggestions to reduce waste at home through personalization according t what residents consume in their own households, enabling them to be better informed about the manner they choose to tackle waste management (Sowers, 1993).

5.5.4 IoT for Waste Management in Smart Homes: Case Study

Example: The Challenge A case that could serve as an example of good deployment of IoT solutions in the smart home community is successful waste management. The project saw smart waste bins equipped with sensors for real-time data collection and connectivity rolled out across the neighbourhood. Autonomous waste-sorting system Oreos: bins that self-sort, measure fill level and communicate with removal services.

By deploying smart waste bins, the amount of waste was reduced significantly & also increased recycling rates. This meant waste management services could manage collection schedules based on real-time data from the bins, fewer trips required to bins and lower fuel consumption. Residents enjoyed a better waste collection service (less missed collections and less mess) and the city gained more savings (Dilrukshi & Kawasaki, 2016).

This was a next-gen project to deliver waste management solutions using IoT. The community was able better waste segregation by incorporating smart technology in daily waste management which led to less operational cost and encouraged a more sustainable operation of their waste management. The case study we walk through showcases how IoT solutions is starting to solve these problems for their time and adding greener into the world.

5.6 AIR QUALITY MONITORING AND ENVIRONMENTAL CONSERVATION

More often than not, these sensors come with mobile apps as well; so you can see and send receive alerts when the air where sensor is placed goes above a safe threshold. This allows for rapid intervention in the form of increased ventilation or use of air purifiers to improve indoor environment quality. Integrating these sensors into broader smart home systems makes a consistent response to air quality issues possible, which is a big plus for folks interested in creating healthier living spaces.

5.6.1 IoT-Based Indoor Air Quality Sensors

Indoor air quality (IAQ) holds a crucial area in social and public effort due to the fact that people spend most of their time indoors, especially at their residence. IoT-enabled indoor air quality sensors help in tracking and controlling IAQ. The smart sensors can identify pollutants including carbon dioxide (CO_2), volatile organic compounds (VOCs), particulate matter (PM2.5), and humidity levels. These devices provide homeowners with up-to-date air quality data to respond to potential health and environmental issues.

5.6.2 IoT-Enabled HVAC Systems for Energy and Air Quality Management

Indoor comfort is closely related to the temperature and air quality of the space in which you are, HVAC systems are used for that. IoT brings further improvements to HVAC systems by adding sensors and connectivity elements to the heating systems which help in making automatic changes based on live data. For example, intelligent thermostats adjust temperature and moisture level based on the current air quality readings automatically balancing energy utilization required to keep a room at setpoint and comfort with the environmental health.

Moreover, HVAC systems IoT tech also helps to leverage predictive maintenance and prevent indoor pollutants from accumulating by checking that the filters and ventilation systems are working properly. These systems can also be programmed to run at energy-saving levels during specific hours, curbing down overall energy costs and lessening the impact on the environment. These air quality monitoring and energy management tactics...support a more sustainable way of living.

5.6.3 Effect of IoT in Indoor Pollution Control

Continuous monitoring and automation are some of the major technologies used by IoT for controlling indoor pollution. Homeowners will be able to determine pollution levels in their house and know what needs mitigated; giving the capability to reduce harmful indoor pollutants with this real-time data based on air quality sensors. When sensors detect high levels of particulate matter or VOCs, for example, the system can trigger air purifiers to turn on automatically or adjust HVAC settings to increase ventilation.

That is, predictive analytics can also help in maintaining pollution through proactive control in IoT-based systems. For instance, it may know from historical data when air quality is likely to go bad (such as more pollution happening outdoors or people in the home engaging in more indoor activity), and change settings before air quality takes a hit. This data-driven method does not just improve indoor air, but also keeps homes abreast of any potential health risks that can arise due to long-term inhalation of indoor pollutants.

5.6.4 Case Study IoT-Enabled Air Quality Monitoring in Households

In a suburban neighbourhood, they implemented an IoT air quality system capable of monitoring the IAQ and taking measures to enhance it. IoT sensors were installed in each home which monitored toxic pollutants (PM2.5 and Formaldehyde), as well as carbon monoxide (CO) and VOCs. Whereas the sensors could connect back to a central dashboard that provided home owners with real-time updates on air quality and send them notifications for potential health perils.

The IoT system was also connected to the homes' HVAC systems, allowing ventilation and air purification to be automatically adjusted according to the sensor data. This led to a significant increase in IAQ and the community was most of the time in a safe zone for pollutants. They also indicated better general health, fewer respiratory problems, and greater comfort from improved temperature and humidity control.

Recent events underscore the power of IoT solutions to address IAQ while showing how real-time monitoring and automation can help create healthier, more sustainable living environments. This accomplishment also demonstrates how IoT-connected home air quality monitors could be mass-deployed to battle poor air quality in homes, especially at a time when clean living and sustainability are proliferating health trends.

5.7 IoT-BASED SMART LIGHTING AND ENERGY SAVING

The big advantage of this automation is that it not only makes your life easier but also has a significant impact on energy efficiency of lighting and avoids using lights when they are not needed. Users can turn on/off lights from afar, eliminating wasteful lighting and unplanned energy consumption for forgotten-on bulbs.

5.7.1 Automated Lighting Systems with IoT

To put it in simple words, IoT encourages every device to submit data points which were non-existent otherwise — e.g., smart radios — the science radio of today saves a significant percentage of what they used to do back in our young memories! IoT-based systems for intelligent lighting development has made its presence felt even more prominently during recent years becoming an essential tool for energy-saving practices within households as well. These systems can automatically control the lights according to situations like daylight availability, room occupancy, and user preferences. Using a central hub or mobile app, homeowners can set lighting schedules, dim and brighten remotely, or configure the system to adapt lights based on real-time data.

5.7.2 AI-IoT Integration for Smart Use of Light

Using AI with IoT-based lighting offers another significant dimension to save more energy. Artificial intelligence-driven algorithms can detect patterns of light usage, anticipate the behaviour of a user and adjust lighting parameters. For instance, the AI system can remember when and where light is used most often in a house to automatically adjust lights accordingly – or simply turn them off when not needed.

When you add AI, the system can predict what your lighting needs are going to be and change light brightness depending on extra weather or sunlight. Sophistication At Its Best and A Reduction in Energy Waste Levels Highly Sophisticated Lighting to Provide the Perfect Amount of Light Here And There With continued use the AI empowered system will become more attuned to how much heating is needed and where, gaining an optimal balance between comfort, aesthetics, and energy efficiency.

5.7.3 Optical Motion Sensing and Light-Level Adaptive Lighting Technologies

The bottom line is that, for any smart lighting system to be truly beneficial in terms of both convenience and energy reduction, motion sensing and adaptive lighting technologies must be included. Capable of identifying movement inside a room, motion sensors trigger lights to be turned on whenever someone enters a zone and turns them off when the chamber is character empty. Thereby illuminating in occupied rooms only, promising an end to the waste that comes with leaving lights on in empty rooms.

Adaptive lighting takes the control outside of the simple on-or-off-of-artificial-light and into a realm where daylight is manipulated without invading the closed space; added artificial light varies in intensity, as do lights automatically selected for colour temperature based upon ambient conditions. During the day, the lighting may dim or brighten when it is cloudy which creates the perfect atmosphere with using almost no energy. These smart adaptations ultimately provide a way to reside at home more comfortably with less power (Dhatterwal et al., 2024).

5.7.4 Smart Lighting in Energy-Efficient Homes

IoT-based smart lighting systems were deployed at multiple homes as part of a project targeting energy-efficient homes for residential sector. Every house was fitted with home automated lighting systems alongside with motion sensors, AI-based control algorithms and the adaptive lightning properties. The lighting systems had been installed as a pilot, and the project monitored how much electricity was being consumed before and after.

The results demonstrated a significant reduction in power consumption with savings of up to 25% on lighting energy appearing at the household level. In the light bulb department, homeowners also boasted of higher satisfaction in terms of automation: auto controls that turn on when you arrive home late at night, and mobile app settings that allow you to change lighting to optimal settings. Also, with time the AI integrated into the systems efficiently recognized user behaviour patterns that helped in channelizing energy use but ensuring a cozy living atmosphere.

The case study is a reminder of what kind of energy efficiency we could achieve in our homes with IoT-based smart lighting systems. Through automation, AI and adaptive technologies—so that they are solved by not only providing environmental solutions and cutting costs but also making lives better for residents.—. This move should give you an idea of what to expect when we start to implement smart lights as a critical instrumentality for energy starred homes in the future.

5.8 DATA COLLECTION, ANALYTICS, AND DECISION-MAKING IN HOUSEHOLD CONSERVATION

With real-time data collection, any anomaly can be attended to by taking prompt action: observing strange water consumption caused by leak, or managing energy use during high-peak. All of this information is in constant motion, so users have more time to analyse and act on the data they receive to minimize consumption and costs.

5.8.1 Real-Time Data Collection Using IoT Networks

IoT enables real-time collection of data from different appliances and equipment in a residential building, delivering insights into how resources are being used by the residents. For example, sensors in appliances, lighting, HVACs and water metres detect minute by minute what consumers are taking from the grid creating massive amounts of data. These data transfer details through wireless communication protocols will help homeowners to monitor and analyse the consumption of electricity, water and waste in real-time (Kaswan et al., 2024).

5.8.2 Big Data Analytics (BDVS) For Household Conservation Monitoring

Since there is continuous and abundant flow of data from IoT devices, big data analytics have a huge role to play in processing, analysing, as well as deciphering the collected information. They utilize machine learning algorithms and sophisticated analytics to analyse how households use resources over time. These tools give homeowners a place to see what their daily behaviours in the home are doing to resource use, and where they can save.

For instance, big data analytics can monitor the overall performance of utilities and alert the homeowner to flow rates or energy consumption that may be out of line with what should be expected from a well-performing system. These analytics tools work to optimize resource usage, evenly distributing schedules across the hours and providing socioeconomic driven insights still trends that help the homes become sustainable through time enjoying reducing cost.

5.8.3 Optimize Household Resource Utilization Using Predictive Analytics

IoT data enables predictive analytics which enables households to predict their ow resource consumption by gauging the usage patterns from past history and considerin external factors such as weather or family schedules. Predictive models forecast futur usage so that proactive steps can be taken; for example, adjusting heating and coolin settings in anticipation of an upcoming heatwave or shifting energy consumption t operate during off-peak hours.

Ideal for detecting leaks and floods before they become a home disaster, this wate leak sensor from Phyn is one of the top smart home devices for staying connected t your plumbing system. With these predictive abilities in their hands, homeowners ca make e fact-based decisions that save energy and limit resource utilization towards cor servation efforts. This includes predictive analytics that can tell when water could b used at a home and makes suggestions on changes of use times to prevent peak demanc improving home efficiency even more.

5.8.4 Household Energy Conservation Using Data-Driven Decision-Making

An initiative for real-time gathering of smart metre and domestic appliances data i residential buildings targeted at saving energy, has leveraged IoT networks in one of it projects. The company analysed this with big data and predictive analytics to undei stand where energy was being wasted and provide actionable insights homeowner could take to save money on their utility bills.

The case study results appeared to show that using data to base your judgement o could reduce energy use by a large amount. Residents could use personalized, real-tim alerts and suggestions to adjust their energy usage as needed. Some homes had signe up to use power at off-peak times, or had swapped out their old appliances for smarte IoT-enabled energy-efficient machines.

Using predictive models, the system predicted energy needs based on weather fore casts and previous data enabling households to make pre-emptive adjustments. As result, during the pilot competition, participants reduced their overall energy consump tion by 20% once data-driven strategies were in place—all helping the average house hold score lower utility bills and a lighter ecological footprint over time.

The case study reinforces how real-time data collection and analytics is essentia to household conservation. IoT networks and data-powered insights enable homeowner to make informed choices, with clear benefits for sustainable living — increasing th likelihood of residential energy management technologies becoming more mainstream

5.8.5 Potential AI-Driven IoT Solutions for Household Resource Optimization

This integration of IoT systems is becoming more enriched with AI to multiply the family conservation efforts on homes. The AI-powered IoT devices are able to perform analysis on huge data that has been generated by household devices, over a period of time learning the patterns of consumption and then makes it easier for them in identifying the areas where iterative gains can be made. AI systems, enabled by machine learning algorithms, can predict consumption patterns and adapt settings as necessary (in real-time) while providing advice to the user on how best to cut down on waste.

For example, AI could automatically direct appliances, regulate lighting according to the availability of natural light or suggest changes in human behaviour to conserve resources. By learning in real-time and continuously improving, AI turns IoT systems more intelligent so that over time they aid households consume less energy, water, food, materials and create less waste.

5.9 CHALLENGES IN IMPLEMENTING IoT FOR HOUSEHOLD CONSERVATION

There are also major structural and financial barriers. While the savings afforded by IoT regulation is significant, most homeowners will find the cost to acquire and install smart metres, sensors, and other equipment up front is too high. Furthermore, the cost of maintaining IoT systems can be a financial strain.

One of the biggest obstacles to realizing IoT for home savings is that such solutions tend to involve a level of technical drone. Reliable hardware such as sensors and actuators, steadfast connectivity thanks to the leviathan which is yes IoT and constant supply of power are imperative for IoT systems to deliver desired results. High-speed Internet, especially in rural areas not available in many homes to support such systems.

5.9.1 Data Privacy and Security Challenges in IoT-Based Homes

IoT gadgets produce a large repository of data linked to your personal habits, purchase patterns, and household behaviour. But that raises questions of privacy and security, if the watches are constantly monitoring data. Failure to safeguard household data could leave it susceptible to cyberattacks gaining unauthorized access or misuse of the information.

Securing IoT Devices and Networks It is crucial to secure systems when it comes to using IoT devices and networks in the IT-world as breached systems can leak data but also trigger threats like device manipulation that can control things such as home appliances from anywhere remotely. The more homes that embrace data-collecting IoT devices to be eco-friendly, the more people must strike a balance between collecting energy-use data and protecting their privacy.

5.9.2 Interoperability Issues with IoT Devices

Another big problem for in-home IoT conservation is interoperability. It is challenging to ensure that devices of different manufacturers cooperate seamlessly with the increasing number of IoT devices. Incompatible Devices, The lack of a unified standardized protocol across devices and platforms can result in compatibility problems, which hinder homeowners from building all-around functional IoT environment.

It might not be possible for a household to effectively manage resources if it has a smart water metre from one company and a smart thermostat from another. The fragmentation of IoT ecosystems may decrease the efficiency of conservation efforts overall and will alienate users.

5.9.3 Regulatory/Standardization Barriers

The more widespread IoT adoption becomes, the sooner it will be necessary to establish regulatory frameworks and other sorts of standardization protocols in order to safeguard the operation of devices in a manner that is safe and secure enough. Nonetheless, the rules for IoT in energy conservation at the home are in flux and standards differ from region to region and industry by industry.

Interoperability, data security, and consumer protection all critically require standardized processes. In the absence of standardized regulation and rating systems, companies designing IoT devices must not develop proprietary systems that does not coordinate with others, hampering a homeowner's ability to achieve complete conservation measures. And they have to do that without putting a halt to innovation and ensuring preservation of data privacy, at the same time promoting adoption of more kind technologies (renewability, for example).

5.9.4 Emerging Trends in IoT for Household Conservation

By connecting these to their homes, people would be able to consume less from non-renewable energy sources and thus reduce the carbon foot print that they are making — thereby also adopting a more sustainable way of living. Furthermore, the systems can provide customers with information on energy production and consumption patterns that allow them to make more informed decisions about how to store and use energy.

5.10 IoT RENEWAL FOR HOME ENERGY SOURCES INTEGRATION

Among the most attractive trends for IoT in household conservation, its connection with renewable energy sources. As the result, connecting IoT technologies with solar panels, wind turbines and others types of renewable energy systems enable households to be more in control of their own power generation and consumption. Reducing the cost of energy produced from renewable sources can be achieved by this technology as IoT-enabled devices monitor and control the production and consumption of energy in real-time. For example, SOLID IoT systems can access solar during peak production hours and divert to stored power or grid energy when solar is low.

5.11 REAL-TIME CONSERVATION MONITORING BY EDGE COMPUTING

Another major trend in IoT for household conservation is edge computing due to its ability to deal with data processing right at the source — like smart metres and different sensors inside households. Edge computing allows you to process data at the device level rather than using cloud-based systems, which is beneficial for real-time analysis and decision-making. This shortens the latency for responding to changes in resource consumption.

An edge-enabled IoT solution will route the water in such a way as to instantly reduce flow of said water should a leak be detected, or even adjust heat and cooling depending on occupancy patterns. Edge computing allows for more rapid decision-making by moving the decision out of the cloud a to a closer proximity to where it is actually occurring in the household IoT systems, thereby enabling higher levels of efficiency and responsiveness that ultimately translate into more effective conservation efforts.

5.12 FUTURE DIRECTION AND RESEARCH AREA

What lies in the future when it comes to IoT for household conservation. A potential area of future work includes the exploration of IoT device costs to develop improved conservation technology for broader use within homes. Also, and with AI and edge computing only going to become more advanced we will see the efficiency in our IoT systems improve further, allowing us to manage resources well into control.

The future directions include two things mainly, one is to integrate IoT with smar grids and commute-based energy systems. The main point of the Internet of Things i that they can help households to share their resources. Innovations in water conserva- tion are also a strong focus for IoT, with many new technologies on the market to help identify leaks more effectively and support automated irrigation in domestic gardens.

Finally, research on data privacy and security will always be a hot topic because the more you use the IoT, the greater your worries are about privacy issues like how user data is collected, stored, and protected. Solving for these challenges will be critical to making IoT-powered consumer conservation solutions viable in the home for the long-term.

5.13 CASE STUDIES OF IoT-ENABLED SUSTAINABLE HOUSEHOLDS

For example, the Freiburg Green City in Germany is a prime illustration of how IoT can achieve economies in an entire region. IoT sensors and automated solar energy, water usage and waste management systems in place in homes under this eco-district. These technologies help reduce the carbon footprint of the Freiburg community as a whole, meanwhile enhancing standards of living.

5.13.1 Sustainable Housing Projects

Real-world housing projects have in recent years benefited from the introduction of IoT solutions to drive sustainability and resource conservation. BedZED (Beddington Zero Energy Development, UK). BedZED incorporates high performance building design alongside IoT-enabled devices to collect and in turn manage energy and water consumption on the fly. The use of smart metres provides near real-time visibility on consumption patterns for occupants and automated control over loads through IoT platforms, as seen in the case of smart thermostats and appliances that change temperature (e.g. heating/cooling) based on occupancy detection and/or external environmental conditions. The initiative shows how IoT could aid in reaching energy neutrality & sustainable living.

5.13.2 Micro Smart Grids and Low Energy/ Green Home IoT Applications

IoT sensors are also making the change in traditional to green and low-energy homes, the critical contribution from automation in IoT technologies. For example, the Sonnenhaus project in Austria used IoT-based solar energy systems to control power usage in homes hat were engineered to be nearly 100% solar-powered. IoT sensors monitoring the gen-ration and storage of solar energy, ensuring consumption patterns align with peak generation hours and surplus energy is captured for use later on Covered boat storage s also offered with power, lighting, heating and cooling all controlled by automated ystems to ensure no or very limited energy waste. While heat houses are the ultimate in efficiency, they illustrate what IoT technology can achieve on a large scale with energy management.

In Project Smart Home, housed COMs in Singapore have been built to save energy and further contributes to the integration of IoT home design. Adaptive lighting systems based on IoT protocols can change depending upon the level of natural light and occu-pancy, cutting down energy wastage considerably. The owners of the firm recommend adopting IoT-integrated HVAC systems that have dynamic responses to the indoor/out-door conditions and then regulate the air quality/temperature of your house in order to meet a more sustainable living standard.

5.13.3 Successful Home IoT Projects

One can learn many interesting takeaways from these successful IoT-enabled sustain-able household projects. Lesson #5: User Engagement and Education are Vital. IoT sys-tems eliminate a lot of the manual work required from homeowners, but they also need to continue to act wisely and change their behaviour accordingly for IoT conservation best practices to have the desired effect. This focus on user-facing interfaces has been a necessity in many projects to push community adoption of sustainable options.

The next lesson is that interoperability matters in IoT systems. Numerous instal-lations have proven the benefits of mixing and matching devices on a single network. They keep the entire home running smoothly at all times, from water saving to energy control. BedZED Example A use case — integration of multi-manufacturer devices via standardized protocols to create a holistic energy conservation ecosystem/Image by Kleindienst Group

Finally, these are case studies in the future — scale beyond a home to neighbour-hood or district. At scale, as demonstrated in the Freiburg and Sonnenhaus projects, the benefits to sustainability and resource preservation are multiplied, aligning their uses with broader environmental and social aspirations.

5.14 CONCLUSION

This presents a major revolution in the aspect of IoT integration as part of efforts in con-servation of households. In our effort towards sustainability, it has dawned on human beings that resource conservation is paramount hence IoT provides distinct solutions to a range of basic necessities such as water, energy and waste. The utilization of smart water metres and integration of leakage detection systems demonstrates how use of real-time monitoring techniques can help minimize wastage of water while on the sam degree use of smart metres and energy-efficient home electrical appliances show ho IoT can be used to minimize electricity usage. But the process of achieving IoT-assiste conservation paradigms is not without obstacles. Factors such as technical constraint funding shortfalls, and deficiencies in the built environment as well as data confident ality issues and system compatibility questions likewise remain a concern so that it crucial to increase the learnings to populations in order to be most fruitful. In the futur the trends including the connection of IoT with renewable energy, edge computatio and AI will unlock new innovation in household conservation. Lastly, IoT becomes solution to support a sustainable living since the households can control the resource they use intelligently. Exploiting the theoretical framework that can be derived fro the present case, it is possible to open the further evolution and development of Io with correct accomplishments laying the foundation for a greener and more sustainab future.

REFERENCES

Alcubilla, R. G., & Lund, J. R. (2006). Derived willingness-to-pay for household water use wit price and probabilistic supply. *Journal of Water Resources Planning and Managemen* *132*(6), 424–433.

Alemu, E. T., Palmer, R. N., Polebitski, A., & Meaker, B. (2011). Decision support system fc optimizing reservoir operations using ensemble streamflow predictions. *Journal of Wate Resources Planning and Management*, *137*(1), 72–82.

Boafo, Y. A., Saito, O., Kato, S., Kamiyama, C., Takeuchi, K., & Nakahara, M. (2016). The role c traditional ecological knowledge in ecosystem services management: the case of four rur communities in Northern Ghana. *International Journal of Biodiversity Science, Ecosyster Services & Management*, *12*(1–2), 24–38.

Dhatterwal, J. S., Kaswan, K. S., Tirapathi Reddy, B., & Gangadevi, E. (2024). Computation intelligence–based image classification for 3D printing: Issues and challenges. In E Gangadevi, M. Lawanya Shri, Rajesh Kumar Dhanaraj & Balamurugan Balusamy (Eds.) *Computational Intelligence in Bioprinting* (pp. 79–91). New Jersey: Wiley and Hoboken

Dilrukshi, R. A. N., & Kawasaki, S. (2016). Plant-derived urease induced sand cementation use in geotechnical engineering applications. In *International Conference on Geomechanics Geo-energy and Geo-resources.*

Fisseha, G., Gebrekidan, H., Kibret, K., Yitaferu, B., & Bedadi, B. (2011). Analysis of land use land cover changes in the Debre-Mewi watershed at the upper catchment of the Blue Nil Basin, North West Ethiopia. *Journal of Biodiversity and Environmental Sciences*, *1*(6) 184–198.

Kaswan, K. S., Dhatterwal, J. S., Batra, R., Balusamy, B., & Gangadevi, E. (2024). 3D Bioprintin technology optimization using machine learning. In *Computational Intelligence i Bioprinting* (pp. 303–321). Hoboken, New Jersey.

May, P. J., Burby, R. J., & Kunreuther, H. (1998). Policy design for earthquake hazard mitiga tion: Lessons from energy conservation, radon reduction, and termite control. *Earthquak Spectra*, *14*(4), 629–650.

Powrie, W., & Beaven, R. P. (1999). Hydraulic properties of household waste and implications for landfills. *Proceedings of the Institution of Civil Engineers-Geotechnical Engineering*, *137*(4), 235–237.

Sønderlund, A. L., Smith, J. R., Hutton, C. J., Kapelan, Z., & Savic, D. (2016). Effectiveness of smart meter-based consumption feedback in curbing household water use: Knowns and unknowns. *Journal of Water Resources Planning and Management*, *142*(12), 04016060.

Sowers, G. F. (1993). Human factors in civil and geotechnical engineering failures. *Journal of Geotechnical Engineering*, *119*(2), 238–256.

Walie, S. D., & Fisseha, G. (2016). Perception of farmers toward physical soil and water conservation structures in Wyebla Watershed, Northwest Ethiopia. *World Journal of Agricultural Sciences*, *12*(1), 57–63.

Xiao, Y., Fang, L., & Hipel, K. W. (2018). Agent-based modeling approach to investigating the impact of water demand management. *Journal of Water Resources Planning and Management*, *144*(3), 04018006.

6 IoT-Based Geotechnical Examining Techniques Incorporating Automatic Practices for Statistics Procurement and Amplification

6.1 INTRODUCTION TO IoT IN GEOTECHNICAL ENGINEERING

The Internet of Things (IoT) is transforming industries left and right, and geotechnical engineering is not immune to it. In the past, this has been accomplished by directly observing and testing soil conditions and underground structures using manual procedures. While these traditional techniques are reliable, they can be less accurate and more time-consuming, and in some cases, not even allow for real-time analysis (Rane

DOI: 10.1201/9781003537625-6

et al., 2024). The advent of IoT is ushering in a new era for geotechnical engineering marked by increased efficiency, safety, and accuracy through data-driven insights and automation across major projects.

Geotechnical engineering is fraught with many challenges, especially because subsurface conditions have high levels of variability, and extracting reliable data over time is difficult. Traditional methods have meant soil stability needed to be monitored, but finding out how this affects the structures above is typically down to costly intervention and periodic assessment of their health (Pal et al., 2021). Environmental issues like changing weather conditions, seismic activities, and land development affect the geotechnical conditions to a great extent, and they make matters even more confusing. To effectively address these challenges, we need innovative solutions to provide around-the-clock real-time monitoring service and robust data acquisition capabilities—enabled by IoT technology (Kulkarni et al., 2022).

These conventional methods have been digitized by using IoT in geotechnical engineering, and their data can now be automatically collected and analysed. Similarly, IoT-enabled sensors can be installed in faraway places to keep track of any changes in soil conditions, slope stability, and other parameters. This enables on-the-spot monitoring of structural changes (Vaish et al., 2021). By maintaining a continuous data stream, manual inspections decrease, and predictions of failures or hazards are more likely to be detected earlier on in geotechnical work rather than waiting as a cause for action.

The major goal of including IoT in geotechnical exams is to make the most of information, reducing acquisition time and enhancing its accuracy while minimizing human interference. IoT systems are built to instantaneously access and assess key structural conditions and provide real-time data for predictive maintenance, safety monitoring, and efficient infrastructure management (Dwivedi et al., 2017).

IoT geotechnical evaluations also provide a significant safety benefit. Automated monitoring enables immediate alerts to the control room, allowing process engineers to take proactive measures before conditions escalate. This can be particularly useful for large-scale construction projects or in areas where natural disasters are frequent. Early warning devices offer alerts that can save lives, while automated IoT systems eliminate the risks associated with manual inspections in hazardous or remotely inaccessible locations, such as underground tunnels and unstable cliffs (Rane et al., 2024).

In addition to safety, the benefits of IoT extend to the sustainability of geotechnical engineering practices. IoT significantly contributes to reducing the environmental footprint by eliminating frequent site visits and manual inspections (Pal et al., 2021). Deploying fewer vehicles and machines to the site benefits the environment in the long term, thus reducing emissions when fewer machines are on the ground, minimizing fuel consumption, lowering noise pollution, and decreasing overall environmental disruption. IoT technologies also help optimize material usage during construction, driven by data-infused insights that allow engineers to make more informed choices on resource distribution (Kulkarni et al., 2022).

Furthermore, IoT enables the integration of other advanced technologies such as big data analytics, cloud computing, and machine learning (ML). The systems can process tremendous amounts of data from IoT sensors to find patterns and trends that would be missed by manual analysis (Vaish et al., 2021). In essence, this allows for more precise predictions regarding future structural behaviour, enabling the design of more resilient and durable structures (Dwivedi et al., 2017).

IoT systems not only predict analytics but also enable real-time geotechnical data monitoring, allowing sensors to run consistently and enabling automated sensor calibrations. This permits prompt corrections of any inaccuracies, maintaining the utmost quality of data integrity (Rane et al., 2024). The amplification and normalization of data in real-time create a far clearer, more realistic view of geotechnical conditions supporting engineers' abilities to make informed decisions based on the most current information (Kulkarni et al., 2022).

6.1.1 Fundamentals of IoT Systems for Geotechnical Applications

This chapter begins with an overview of the basics of IoT systems, especially in geotechnical applications, such as main components, data acquisition methods, communication protocols, and development processes that integrate into traditional geotechnical practices. These all contribute to how effectively and efficiently geotechnical conditions can be monitored with geological data management.

6.1.2 Components of IoT and Architecture

There are a few basic components of IoT systems: sensors, communication networks, and data processing units. The sensors are distributed in the field to monitor the environment (for example, soil content, temperature, or load). These sensors are linked to data acquisition systems, which serve to process and store the collected data. Data transfer from sensors to the central system is carried out using communication networks such as wireless and cellular systems. An IoT system architecture consists of edge devices with initial data processing, cloud platforms that serve as the backend for mass data storage, plus advanced analytics and user interfaces to deliver useful insights and alerts to engineers (Hong et al., 2011).

6.1.3 Sensors and Data Acquisition for Monitoring in Geotechnical Engineering

Because sensors are what provide the necessary data for monitoring subsurface and structural conditions, they are crucial to every geotechnical IoT system. At the same time, other sensors are frequently used, such as strain gauges for deformation measurement, piezometers for measuring water pressure, and accelerometers for assessing vibration, along with inclinometers for monitoring slope stability, LIDAR for surface mapping, and temperature sensors for thermal analysis. Data collection is where each sensor reading is collected, converted, and stored. Modern IoT systems use various sensor technologies available, presenting high accuracy and reliability. These sensors are usually part of a networked set of devices collecting data on an ongoing basis to allow for far-reaching observations of the geological internment condition (Firoozi et al., 2024).

6.1.4 IoT Wireless Communication Protocols

Wireless communication protocols are vital for the proper functioning of IoT systems because they help devices transmit data and communicate with each other without any physical connection. Geotechnical applications often prefer to use Wi-Fi, Bluetooth, or cellular networks, including 4G and 5G, each having its own sets of benefits: Wi-Fi can provide high data rates; Bluetooth is useful for short-range communication with low power consumption; finally, cellular networks cover a broad area enabling geotechnical deployment in areas where it was previously impossible. In addition, low-power wide-area networks (LPWAN), such as LoRaWAN, are now commonly implemented because of their long-range and low-power features, which play to the needs of large-area geotechnical monitoring systems (Abbasnejad et al., 2024) (Table 6.1).

6.1.5 IoT along with Geotechnical Practices

Combining IoT with conventional geotechnical practices includes setting new technologies in general as per the traditional techniques for better investigation and monitoring. Bringing real-time data collection and analysis together gives faster views on geotechnical conditions, providing a timelier assurance. For example, embedding IoT sensors into boreholes, footings, or retaining walls for 24/7 monitoring with no interruption to business as usual allows for the complementarity of IoT data with manual measurements, resulting in a much better and more complete picture of the geotechnical conditions that supports decision-making (Broekman et al., 2020). The marriage of IoT and traditional methods provides perfect accuracy as well as more efficiency and safety in geotechnical projects.

6.1.6 Automated Practices for Geotechnical Data Procurement

The pairing of automated methods in geotechnical data acquisition is evolving the way we capture, analyse, and interpret data. They improve the accuracy of calculations, reduce time, and thus help manage complex geotechnical projects. Below is a high-level view of the shape of automation and data acquisition within this domain.

Automated Deployment: Use robotics and advanced equipment necessary to release the sensors on their own at the exact spot where they need to be placed. By using this

TABLE 6.1 IoT wireless communication protocols

PROTOCOL	*RANGE*	*DATA RATE*	*POWER CONSUMPTION*	*APPLICATION AREAS*	*NETWORK TOPOLOGY*
Zigbee	10–100 m	Up to 250 kbps	Low	Home automation, smart lighting, energy monitoring	Mesh, Star
Wi-Fi (802.11n/ac)	50–100 m (indoor)	Up to 600 Mbps (802.11n), 1.3 Gbps+ (802.11ac)	High	Smart home devices, video streaming, internet-based IoT applications	Star, Point-to-Multipoint
Bluetooth low energy (BLE)	10–50 m	Up to 2 Mbps	Very low	Wearables, medical devices, short-range IoT sensors	Star, Mesh
LoRaWAN	2–15 km (rural), 2–5 km (urban)	0.3–50 kbps	Very low	Long-range sensor networks, agriculture, smart cities	Star
NB-IoT (narrowband IoT)	Up to 35 km	Up to 250 kbps	Low to very low	Smart metering, environmental monitoring, industrial IoT	Star
Sigfox	10–50 km (rural), 3–10 km (urban)	100 to 600 bps	Very low	Low data rate applications, asset tracking, smart metering	Star
Z-Wave	30–100 m	Up to 100 kbps	Low	Home automation, security systems, lighting control	Mesh
LTE-M (Cat-M1)	Up to 11 km	375 kbps–1 Mbps	Low	Asset tracking, industrial IoT, smart transportation	Star
6LoWPAN (IPv6 over low-power wireless personal area networks)	10–100 m	Up to 250 kbps	Low	Smart grid, industrial automation, environmental monitoring	Mesh, Star
Thread	30–50 m	Up to 250 kbps	Low	Home automation, smart lighting, security systems	Mesh

technology, it is guaranteed that sensors are installed accurately and consistently, which reduces human error and greatly enhances data quality (Jacobsz et al., 2004). Calibration systems are also critical; they are required to be done regularly to ensure accuracy and reliability, and an automated system allows for calibrating sensors over time with minimal human intervention, improving efficiency and consistency. This enables hands-free automation of calibration checks and adjustments on a scheduled basis to ensure that data remains trustworthy over its lifetime. This kind of automation is particularly useful for large-scale projects where manual calibration would be far from efficient.

6.1.7 Real-Time Data Collection

Effective geotechnical monitoring requires a provision for real-time data collection so that the engineers receive the output and can analyse it as data is generated. In our day and age, monitoring has become an active process through sophisticated IoT systems, more commonly represented as sending sensor data to a centralized database or cloud platform. Streaming data from sensors to real-time dashboards allows geotechnical conditions to be monitored in real-time (Chawla et al., 2020). It is important to catch any problems with soil stability or structural issues early so that we do not have a bigger issue on our hands later, while also enabling real-time data collection for engineers to use the most up-to-date information when making decisions.

6.1.8 Remote Data Transmission with WSNs

Geotechnical monitoring systems often incorporate wireless sensor networks (WSNs) due to their utility in remote sites and difficult-to-access sites. These networks consist of a series of sensors connected together that communicate wirelessly to send data back to a central system. Zigbee, LoRaWAN, and cellular networks are most often used for their range and reliability. WSNs minimize the requirement for large-scale cabling and infrastructure, thereby facilitating installation and management. Specifically suited to applications in which sensors are spaced at great distances from one another—like monitoring landslides or expansive construction projects—these systems enable access to critical data otherwise not transmitted elsewhere. Data capture is possible from nearly any location due to the wireless networks, which provide control and analysis for a complete gamut of geotechnical monitoring (Villa et al., 2021).

6.1.9 Geotechnical Automated Data Acquisition

A number of real-world case studies demonstrate how automated data procurement in geotechnical engineering leads to direct benefits. In a major dam construction project, for example, automated sensors were emplaced to continuously monitor soil pressure and structural strain. The sensors were kept true throughout the project using automated calibration systems. Engineers could collect vital data in real-time, allowing them to anticipate any potential problems early and make timely adjustments to avoid issues and improve safety. WSNs facilitated round-the-clock monitoring of ground conditions and immediate responses to any indications of instability during a large-scale tunnel construction job. These case studies detail practical experience in how automation helps speed up data acquisition and ensures enhanced workflow safety and management.

6.2 AMPLIFICATION OF GEOTECHNICAL DATA USING IoT TECHNOLOGIES

IoT technology helps to amplify geotechnical data while also improving the accuracy and utility of information collected from different sensors and monitoring systems. Using high-quality data processing and analysis tools, geotechnical engineers can access a much more detailed picture of the subsurface or structural conditions than ever before, which enables them to make better informed choices at each stage of their project. This four-part blog series provides a deeper dive into how data amplification works and where the difficulties arise.

$$4(t) = \sum_{i=1}^{N} \left(\alpha_i \cdot \frac{G_i(t)}{S_i(t)} + \beta_i \cdot \frac{H_i(t)}{C_i(t)} \cdot \log\left(1 + \frac{D_i(t)}{R_i(t)}\right)\right)$$

$$G_i(t) = \frac{\int_0^T \frac{\partial P_i(t)}{\partial t} \cdot W_i(t)\,dt}{\int_0^T W_i(t)\,dt}$$

$$S_i(t) = \frac{\partial^2 P_i(t)}{\partial t^2} + \lambda \cdot \frac{\partial^3 P_i(t)}{\partial t^3}$$

$$H_i(t) = \frac{\int_0^T \frac{\partial^2 E_i(t)}{\partial t^2} \cdot P_i(t)\,dt}{\int_0^T \left(1 + \log\left(1 + \frac{\partial E_i(t)}{\partial t}\right)\right) dt}$$

$$C_i(t) = \frac{\partial}{\partial t}\left(E_i(t) + \mu \cdot \frac{\partial^2 E_i(t)}{\partial t^2}\right)$$

$$D_i(t) = \frac{\sum_{j=1}^{M} \log\left(1 + \frac{P_{ij}(t)}{E_{ij}(t)}\right)}{\int_0^T \frac{\partial^2 P_{ij}(t)}{\partial t^2}\,dt}$$

$$R_i(t) = \int_0^T \frac{\partial P_i(t)}{\partial t} \cdot \log(1 + E_i(t))\, dt$$

$$\frac{dA(t)}{dt} = \sum_{i=1}^{N} \left(\alpha_i \cdot \frac{d}{dt}\left(\frac{G_i(t)}{S_i(t)}\right) + \beta_i \cdot \frac{d}{dt}\left(\frac{H_i(t)}{C_i(t)} \cdot \log\left(1 + \frac{D_i(t)}{R_i(t)}\right)\right)\right)$$

$$\frac{dG_i(t)}{dt} = \frac{d}{dt}\left(\frac{\int_0^T \frac{\partial P_i(t)}{\partial t} \cdot W_i(t)\, dt}{\int_0^T W_i(t)\, dt}\right)$$

$$\frac{dS_i(t)}{dt} = \frac{\partial}{\partial t}\left(\frac{\partial^2 P_i(t)}{\partial t^2} + \lambda \cdot \frac{\partial^3 P_i(t)}{\partial t^3}\right)$$

$$\frac{dH_i(t)}{dt} = \frac{d}{dt}\left(\frac{\int_0^T \frac{\partial^2 E_i(t)}{\partial t^2} \cdot P_i(t)\, dt}{\int_0^T \left(1 + \log\left(1 + \frac{\partial E_i(t)}{\partial t}\right)\right) dt}\right)$$

$$\frac{dC_i(t)}{dt} = \frac{d}{dt}\left(\frac{\partial}{\partial t}\left(E_i(t) + \mu \cdot \frac{\partial^2 E_i(t)}{\partial t^2}\right)\right)$$

$$\frac{dD_i(t)}{dt} = \frac{d}{dt}\left(\frac{\sum_{j=1}^{M} \log\left(1 + \frac{P_{ij}(t)}{E_{ij}(t)}\right)}{\int_0^T \frac{\partial^2 P_{ij}(t)}{\partial t^2}\, dt}\right)$$

$$\frac{dR_i(t)}{dt} = \frac{d}{dt}\left(\int_0^T \frac{\partial P_i(t)}{\partial t} \cdot \log(1 + E_i(t))\, dt\right)$$

Symbols Explanation:

- $A(t)$ is the amplification factor at time t.
- $G_i(t)$ is the data gain function for sensor i over time.
- $S_i(t)$ is the smoothing factor of data for sensor i at time t.
- $H_i(t)$ is the historical trend function of the sensor i data.
- $C_i(t)$ is the correction factor for geotechnical data for sensor i.
- $D_i(t)$ is the displacement parameter affecting the amplification for sensor i.
- $R_i(t)$ is the residual data impact over time for sensor i.
- $P_i(t), E_i(t)$ is the geotechnical pressure and environmental influence readings for sensor i at time t.

- α_i, β_i are the sensor-specific weighting factors.
- λ, μ are the regularization constants.
- $W_i(t)$ is the weighting function dependent on environmental variables.
- M, N are the total number of sensors and data points, respectively.

6.2.1 Enhancing Accuracy with Data Amplification Techniques

Data Amplification: Improve and increase the quality and accuracy of data acquired from geotechnical sensors. It has acquired specialized services such as advanced statistical methods, signal processing, and calibration. One example would be that filtering algorithms could be used to filter noise from sensor data so that the measurements actually reflect g-layer (geotechnical layer) conditions. Furthermore, methods where data fusion aggregates the information collected by different sensors to provide a comprehensive and precise understanding of the observed system are considered powerful tools for enhancing decision-making, improving accuracy, and optimizing system performance. When we do that, it makes analysis more reliable and gets our engineers closer to the truth.

6.2.2 Geotechnical IoT Data Preprocessing

For reliable data on geotechnical IoT applications, it is paramount to normalize and filter the data. Normalization is performed to scale values so that they fall into the same range, which allows them to be more easily analysed and compared. For instance, data collected by various sensors that sense the same measurable parameter will be normalized due to different measurement units or scales. This is where filtering techniques, for example, moving average or median filters, are applied to smooth out the fluctuations and eliminate noise from sensor readings [16]. These processes ensure the creation of better and cleaner datasets which can now undergo additional analysis and further decision-making.

6.2.3 The Future of Data Amplification Using AI and ML

ML and AI as predictive data magnifiers have expanded to geotechnical engineering too. AI and ML algorithms search for patterns in enormous quantities of data that are not identifiable through common means. For the design lecturer, another goal is to use predictive models of geotechnical conditions for forecasting based on historical data and existing information. For instance, ML models can forecast potential soil instability or structural failures and by consuming past incidents and live sensor readings. This eature significantly enhances the ability to predict and control hazards that may harm a worker, making geotechnical projects safer and more successful.

There are many challenges that arise with data amplification in geotechnical applications, even though there are benefits. Chief among them are the difficulties involved n consolidating data from various sources with disparate formats, making normalzation and analysis a real challenge. In addition, sensor inaccuracies and calibration ssues would affect the quality of the data, thus causing errors when these results are amplified. To level up to address these challenges, robust calibration protocols and more advanced filtering techniques are needed that allow for better EMG data management. These may be obviated by automated systems for real-time calibration and standard data formats. Besides, ongoing AI and ML validation against actual data is essential to est the functionality of AI strategies and ensure that models are accurate, reliable, and capable of making informed decisions in real-world applications.

6.3 ADVANCED IoT-BASED TECHNIQUES FOR GEOTECHNICAL EXAMINATIONS

The use of modern IoT-enabled technologies in geotechnical surveys has transformed the process of monitoring and interpreting geomaterial status of soil, structures, or landforms. They can access real-time data, predictive analysis (using ML and algorithms), and remote monitoring, making our ability to manage and mitigate risks on geotechnical projects much more effective.

6.3.1 Soil Real-Time Monitoring and below Ground Analysis

Real-Time Soil Monitoring: It refers to monitoring soil conditions continuously by IoT-based sensors installed in the field on a regular basis. It can measure important parameters such as soil moisture, temperature, and pressure and give immediate feedback on the behaviour of the soil. With subsurface analysis, the use of ground-penetrating radar (GPR) and electrical resistivity tomography (ERT) can pull up granular images of subsurface conditions. This information is also key in determining soil stability and recognizing significant changes that could point to potential problems, such as erosion or settlement. These strategies also offer real-time insights that help engineers make decisions and act fast enough to stabilize and make the projects safe.

6.3.2 IoT in Structural Health Monitoring of a Foundation

Here, IoT technologies serve as an essential enabler for monitoring the structural heal of foundations (SHM). Parameters like load distribution, vibration, and deformatic are tested by the sensors embedded in motions and structures. The data is then sent a centralized system where it is processed and the status of the foundation is designe In the best systems, alerts that something is going wrong or starting to fail are sent system engineers with enough information about what could go wrong and how like it is that it will occur. Electrical resistance strain gauges and accelerometers can als monitor the impact of loading as well environmental conditions on foundation stabilit which are useful indicators for long-term structural integrity.

6.3.3 Smart IoT Systems for Landslide and Earthquake Prediction

When it comes to predicting natural phenomena like landslides and earthquakes, sma IoT systems are exploring their applications. These systems work by using a networ of sensors that monitor ambient conditions sensitive to the events, like changes in so moisture, ground deformation, or seismicity. ML: These algorithms are designed to us collected data, analyse, find patterns, and predict possible occurrences. In areas suc as landslide-prone regions, sensors placed in the soil can measure changes in moistur and movement of the ground to monitor for risks. Correspondingly, seismometers an accelerometers can monitor seismic events to warn of an imminent earthquake or th failure of a nuclear structure.

IoT devices are capable of monitoring slopes and embankments constantly, whic is why their application significantly improves slope stability monitoring over lon distance. These devices are inclinometers, piezometers, and remote sensing technolo gies measuring slope angle, soil pressure, or groundwater levels. This collected dat is then transmitted wirelessly to central monitoring stations, where it can be analyse and reported in real-time. Monitoring slopes from a distance allows engineers to catc warning signs of instability or movement, which in turn can prevent potential slop failure. It especially comes in handy in areas where it is not possible or extremely dan gerous to send personnel using traditional monitoring techniques.

6.4 AUTOMATED DATA PROCESSING AND ANALYTICS

The use of automated data processing and analytics has become an important component of geotechnical engineering—and modern software solutions facilitate the management, analysis, and visualization of increasingly complex datasets. Advanced visualization, cloud-based solutions, big data analytics, and edge computing together deliver deeper insights and help geotechnical projects to make more informed decisions.

6.4.1 Geotechnical Data Storage and Management in the Cloud

Cloud-based solutions have transformed the landscape of geotechnical data storage and management, providing scale-able and flexible platforms to deal with large volumes. These solutions enable a centralized storage of data from diverse sensors and monitoring systems, locatable through the internet. A Cloud platforms offer a wide range of data storage, backup, and recovery choices, which help ensure the integrity and security of your data. They also allow for shared analysis, meaning many stakeholders can look at and manipulate the data together at the same time. Moreover, it also improves data management performance and minimizes the capacity requirement on-premises besides providing frictionless data linking among various geotechnical projects.

6.4.2 Big Data Analytics for Geotechnical Insights

Big data analytics is an essential process that helps to mine valuable information from the massive amount of collected data generated from geotechnical monitoring systems. ML and statistical modelling are more advanced analytical methods that can be used with massively large data to point out trends, patterns, and anomalies. For instance, the implementation of big data analysis aids in identifying correlations between the properties of soil and physical configurations, which allows for predicting potential failures and optimizing maintenance timetables. With these methodologies, geotechnical engineers can make informed decisions that drive better project results and improved safety and efficiency. This makes it possible to process and interpret huge datasets, which allows for a much better appreciation of geotechnical conditions and helps manage them more proactively.

6.4.3 IoT—Edge Computing in On-Site Data Processing

With edge computing, data processing happens closer to where the data is generated, leading to lower latency, which means the data is processed faster and handled more efficiently. For geotechnical use cases, edge computing means deploying processing capabilities in the field and proximal to where we gather data through sensors and monitoring devices. This approach enables live data, integrated into most of the other important components, minimizing interaction with central systems. For example, if you are processing data from an array of sensors to pinpoint variances or launch alerts, it is likely too urgent for you to wait for batch processing, requiring real-time or near-instantaneous analysis instead. It therefore reduces the load on network bandwidth and increases the overall effectiveness of geotechnical monitoring systems by handling the data locally.

6.4.4 Geotechnical Data Visualization and Reporting

To effectively interpret and communicate all this geotechnical data, effective data visualization and reporting are vital. This is where newer visualization tools and techniques such as visualizations of complex 3D models, geospatial mapping, and interactive dashboards help engineers present data in an easy-to-visualize format. They extract data from multiple sources to present a summary of patterns, trends, and potential issues with visual rendering tools. For visualization tools, reporting features are available to give you a detailed report and summary where findings can be shared with stakeholders for decision-making. Working together to ensure clear and accessible geotechnical data visualization, visualization and reporting tools are valuable in ensuring informed decision-making as well as enhanced communication of the project results.

6.5 INTEGRATION OF IoT WITH TRADITIONAL GEOTECHNICAL METHODS

The marriage of IoT and traditional geotechnical methods is a substantial step forward for the practice which successfully synergizes the advantages of both time-tested principles and real-time data solutions. This integration helps increase the efficacy of your geotechnical monitoring, ensuring data quality and more informed decision-making. Five ways IoT integration is changing traditional geotechnical methods are shown in Figure 6.1.

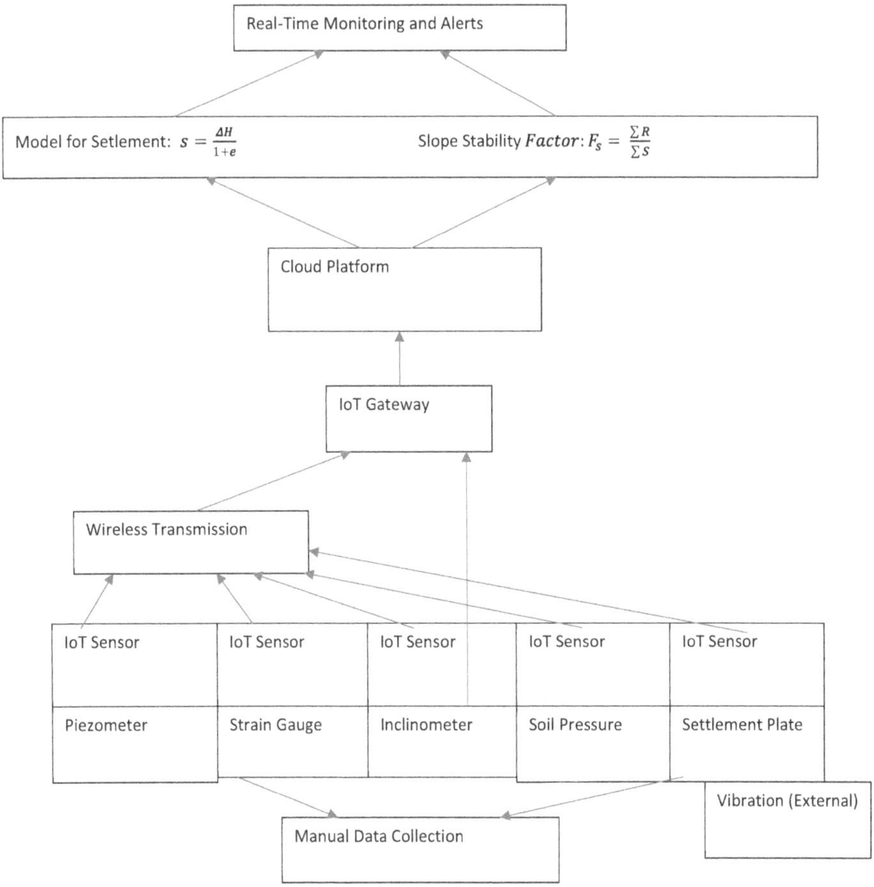

FIGURE 6.1 Integration of IoT with traditional geotechnical methods.

6.5.1 Integrating Manual and Automated Monitoring Techniques

By combining manual with automated monitoring practices, a full geotechnical assessment can be undertaken. Visual inspections and manual readings are also used to provide detailed, hands-on insights into geotechnical conditions through traditional manual methods. These techniques, however, are also time-consuming and restricted. Engineers can leverage automated IoT systems like sensors and data loggers to supplement manual observations with continuous real-time data gathering. This provides complete coverage, improves the quality of the data captured, and makes assessments more frequent. This means that although manual inspections would be able to identify what damage must be remedied, IoT sensors can operate round-the-clock examining changes and making notifications for further investigation.

6.5.2 Improved Borehole Testing through IoT Sensors

While borehole testing has, for years, been the traditional means to complete such tests IoT sensors can bring an additional layer of security through continuous and real-time monitoring of subsurface conditions. In order to obtain soil and rock samples for laboratory tests, the traditional method was drilling boreholes which are often time-consuming and less reliable as the number of data points is very few. Using IoT sensors in boreholes, engineering companies can monitor several vital parameters like soil pressure, moisture levels, and temperature in real-time. This near-constant data collection helps provide a more complete picture of the subsurface and results in better predictions on how the soil will behave. Pressure transducers, for example, provide accurate measurements of pore pressure over time, providing useful information on both groundwater dynamics and soil stability.

6.5.3 Hybrid Case Study in the Most Critical Projects

Take as an example the integration of IoT with traditional geotechnical means into a real-life situation in a major infrastructure project, consisting of building a bridge over seismic land. Standard geotechnical methods (e.g., borehole drilling and soil excursions) were flanked by IoT technologies like real-time seismic sensors and fully automated soil moisture monitors. These manual methods provided critical base data and knowledge of the soils, whereas the IoT sensors could provide constant observation and predictive ability. The changes in soil conditions and structural performance could be monitored over time to ensure the bridge remained safe and reliable. In this way, the integration resulted in not only less reactive intervention but also quicker responses to incidents representing another demonstration of combining traditional with the most advanced technologies.

6.6 GEOTECHNICAL SAFETY AND RISK MANAGEMENT USING IoT

The application of an integrated system is particularly useful in the geotechnical safety and risk management process, as IoT technologies have enabled early warning systems to be triggered, real-time assessment of risk situations, and remote monitoring of high-risk infrastructures. This will allow for better geohazard management and ensure an enhanced level of safety in geotechnical works.

6.6.1 IoT-Driven Early Warning Approach to Geohazards

IoT-enabled early warning systems are important to minimize the risk of geohazards like landslides, floods, and earthquakes. In this context, these systems are based on a network of sensor nodes installed in risky sites to continuously monitor environmental data that may indicate the incidence of geohazard events. Soil moisture sensors, for example, can sense shifts in the Earth's humidity that indicate an impending landslide, and seismic sensors track ground vibrations that suggest an earthquake is imminent. It analyses the data gathered in real-time to detect early-stage warning signs and raise alarms. These systems give early warnings so that evacuations can be made in time and other steps can be taken to avoid loss of life and property.

6.6.2 IoT for Real-Time Risk Assessment

Accept: Continuous monitoring and analysis of geotechnical conditions create a real-time risk assessment through IoT technologies. These include sensors for IoT that offer real-time monitoring of ground shaking, soil stability, and how the structure is holding up. But more importantly, it allows advanced analytics platforms to process this data so that the current risk levels can be assessed and potential issues can be predicted. Ground movements detectable in real-time could represent the precursor stability loss (e.g., slip surface formation) on unstable slopes that ultimately lead to landslide occurrence, while strain and deformation signals indicate structural stress beyond safe tolerances. Engineers use this ongoing risk assessment to complement their decisions and decide how they can address risks on time.

6.6.3 Safety through Remote Monitoring of Critical Infrastructure

IoT technology is used to remotely monitor critical infrastructure, and through regular surveillance, structures (dams, bridges, and tunnels) are monitored continuously. These structures are equipped with IoT sensors, which are used to sense the data of parameters such as load, vibration, and environment. The data is sent to central monitoring systems, which can analyse it and assess the safety and performance of the infrastructure. With remote monitoring, engineers can identify problems early on and assess the condition which allows them to carry out preventative maintenance activities that prevent minor issues from becoming major. This is incredibly useful in areas where a site inspection might prove impractical, or in some cases, too dangerous.

6.6.4 Case Studies—IoT for Disaster Risk Reduction Applications

Use case studies demonstrating the efficacy of IoT solutions for disaster risk manage ment. This is the case, for example, with IoT-based monitoring of structures to asses structural health following seismic events encountered in the field. For example, in a cit that is subject to earthquakes, they deployed a smart monitoring system to constantl monitor the state of buildings. For example, IoT sensors that measure water levels an soil saturation in places prone to flooding have helped avoid damage due to floods b enabling early warnings and effective and quick response. The case studies below illus trate the tangible advantages of IoT solutions for disaster readiness and response.

6.7 STANDARDS, PROTOCOLS, AND BEST PRACTICES

Widespread adoption of IoT in geotechnical engineering requires the formulation c regulatory standards, data security protocols, and best practices to guarantee the rel ability, security, and effectiveness of IoT-based solutions. These guidelines are crucia to standardize the processes and protect the integrity of geotechnical data during col lection, transmission, and analysis. Regulatory standards are critical in defining the saf and effective operation of IoT in geotechnical engineering.

The rationale underpinning the guidelines is to specify the conditions under whic IoT devices and systems should operate to comply with the safety, operation, and envi ronmental considerations. In many parts of the geotechnical world, local and globa engineering guidelines limit the use of geotechnical IoT to establish sensors, data trans mission structures, and systems. For instance, the ISO and IEEE set standards for th integration of IoT devices in specific engineering conditions to ensure that systems, a well as the sensors, can harmoniously operate in geotechnical environments. Ensurin the IoT systems in geotechnical projects align with compliance will guarantee th safety, efficiency, and accuracy of the systems.

Protocols to guarantee data security and privacy in IoT systems: Since colos sal amounts of critical information are collected using the IoT system in geotechni cal engineering, data security and privacy remain central challenges. Strong protocol are essential to prevent unauthorized data access and breaching. One such protocol i encryption which secures the data during transmission and ensures that only the autho rized parties can access and decode the information. Access control mechanisms such as the authentication and authorization protocols limit access to protected informatio by various persons. Privacy protocols are crucial more so in projects whose data col lected may expose sensitive infrastructural vulnerabilities.

Compliance with privacy regulations, such as the general data protection regulation, is essential for the security and data integrity of IoT systems. Best practices for installing IoT geotechnical solutions: Installing IoT-based geotechnical solution requires defined planning and adherence to the best practices to amplify the system's effectiveness and life span; first, the site assessment should be conducted before the sensor installation to understand the best place to put them for data accuracy.

The sensors should be maintained and calibrated regularly to ensure the data maintains its accuracy over time. The sensors need a scalable system design to allow for future expansions, bearing in mind the changing nature of the project. IoT system should have a design that allows them to use minimal energy due to the need to evaluate the efficiency of energy for remote areas where batteries for sensors are used. The data management framework installed should be robust in real-time processing and able to analyse the stored data for fast decision-making. Lastly, the engineers and technicians must maintain continuous training to ensure the use of the IoT systems.

6.8 CASE STUDIES: REAL-WORLD APPLICATIONS OF IoT IN GEOTECHNICAL ENGINEERING

Applying IoT technologies to geotechnical engineering significantly delivers improvement in monitoring, management, and safety for even the most challenging applications. Case studies: examples of IoT in practice. The following are examples of how IoT benefits in real life, and they provide a frame of reference for how beneficial it is as well as ways you might be able to use it.

6.8.1 IoT for Dam/Levee Monitoring

Dams and levees are essential infrastructures that must be monitored constantly to maintain their integrity. However, IoT technologies have been leveraged efficiently to improve the monitoring of this. For example, sensors installed in dam components like water levels, pore pressure, or even structural strain can provide municipalities with real-time data. For example, the Oroville Dam in California used IoT sensors to monitor the spillways and foundations of the dam. It collected real-time data which helped in the early detection of probable issues like seepage or structural weakness. As a result, unplanned inspections could be undertaken in exceptional circumstances and preventive measures taken, which have greatly enhanced the safety and reliability of the dam. With an IoT-based monitoring system, feedback was available in real-time, and engineers used this info to take necessary actions in response to any glitches.

6.8.2 Smart Foundations: IoT in Urban Construction Projects

Urban construction projects commonly require the support of geotechnical professionals, especially when completed in urban areas. IoT technologies power smart foundations, which play a growing role in making urban structures more stable and secure. For example, in a large personal development planned for a city, IoT sensors were positioned inside the cement foundations of a brand-new skyscraper. Soil conditions, load distribution, and structural stress were continuously monitored by these sensors. The data gave engineers useful knowledge about how the building interacts with the land beneath it, helping them to come up with more efficient foundation designs and better ways of building. Real-time post-construction monitoring was also enabled by this innovative solution, so any issues at a later stage can be addressed on time. It was an example of the integration of sensors in urban infrastructures that can improve both construction safety and performance through the deployment of IoT technologies.

6.8.3 IoT Applications in Tunnel and Underground Structure Monitoring

Tunnels and other underground infrastructure are complex due to the limited access to monitoring activities and because geotechnical conditions need to be taken into account. These problems have been effectively dealt with using IoT technologies. In the construction of the Crossrail project in London, for instance, IoT sensors were implemented into the tunnel network to monitor ground movement, groundwater levels, and tunnel stability. These sensors collected real-time data that were used to monitor the state of the tunnel and provide seamless detection of any potential trouble. Not only does this allow early-stage soil stabilization issues to arise which would enable them to get rectified during earthworks but also help in managing how the next stages of construction should be done with a neutralization or avoidance of ground subsidence, leading to safety precautions both on-site and to its adjoining areas. This project was successful at bridging the gap between some of the challenges that come from underground infrastructure and the power of IoT.

6.8.4 IoT-Based Geotechnical Surveying in High-Risk Zones

High-risk zones, such as earthquake-prone areas or landslide locations, receive great benefits from geotechnical examination using IoT technologies. One such notable instance is the application of IoT systems to monitor slope stability in a disaster-prone landslide site in Japan. This required the deployment of a network of sensors to measure soil moisture, slope angles, and ground movement. The sensors collected real-time data, which was used to gauge the danger of a landslide and give digital alerts to nearby authorities and residents. With 24x7 monitoring and alerts, the IoT-driven system helped them stay prepared for any disaster situation that may arise. Case study—How IoT can help in managing geotechnical risks in high-risk zones, resulting in reduction of natural hazards impact and intermediate life loss.

6.9 FUTURE TRENDS IN IoT-BASED GEOTECHNICAL ENGINEERING

Geotechnical engineering is changing fast as well, using IoT technologies. But in the future, many new technologies and trends can take IoT technology to the next level that integrates these existing works done around this domain more comprehensively and expands its possibilities in geotechnical engineering.

6.9.1 Innovations and Upcoming Technologies

Several emerging technologies and innovations will shape the future of IoT in geotechnical engineering. Technological advances in sensors are enabling the creation of new, more accurate, and sturdier sensors that go beyond measuring a single geotechnical parameter. Meanwhile, new developing forms of communication technology in 5G and similar technologies are set to change the way data is transmitted with faster, more stable, and greater bandwidth connections. This will facilitate real-time data streaming from places further away and demand-blocking ones with less lag. Moreover, energy harvesting techniques such as solar-powered sensors are expected to contribute to increased sustainability and life expectancy of IoT devices for geotechnical applications. These advancements will, of course, use precision, accuracy, and scalability on the geotechnical scale.

6.9.2 AI, ML, and Blockchain in Geotechnical IoT

IoT-based geotechnical systems are also evolving and gaining more traction because of AI and ML. By crunching even more data, applying AI and ML algorithms can enable us to discover geotechnical patterns—potential liabilities—as well as anticipate issues and optimize system performance. To illustrate, predictive maintenance models predict the failure of a piece of equipment before it happens, thus preventing its downtime and expensive repairs. Moreover, AI-powered analytics step up the data interpretation, enabling a more advanced understanding of soil behaviour, structural health, and environmental conditions.

IoT-based geotechnical engineering would benefit from blockchain technology as well. Thus, with blockchain, an additional layer of trust through the public ledger can ensure that the geotechnical data is stored securely and verified without manipulation preserving the integrity and transparency of the data. Usher in the age of blockchain-secured and tamper-proof record-keeping for that ever-so-critical geotechnical data—think sensor readings or construction logs—ensuring data integrity, transparency, and seamless access for engineers, regulators, and project stakeholders. It guarantees that all parties involved only use verified information, which is critical to geotechnical projects for conformity and decision-making.

6.9.3 Prospects in Geotechnical Examination and Observation

The future of geotechnical investigation and monitoring will be in integrated and automatic systems. As the IoT comes together with robust data analytics and visualization tools, it will give birth to fully featured monitoring platforms that generate real-time insights along with predictive analytics. These will allow geotechnical engineers to conduct routine reviews and data-based evaluations with improved precision.

On top of this, IoT combined with other technologies, including drones and robotics, will open a range of additional opportunities for geotechnical survey and inspection. Such drones can fly over construction sites or cover areas that are hard to reach—as the IoT sensors they carry will ensure there is a clear picture of every inch of your project—while robotic systems could conduct thorough inspections of underground infrastructure to an extent human employees would need days.

These innovations will be the focus of geotechnical engineering as it continues to embolden each other and share data between different players in the ecosystem. It will ensure that IoT systems are interoperable across platforms and organizations in the adoption of standard data formats, allowing seamless data integration and communication. This will drive better management in geotechnical projects, thereby providing new assets that are safer to build and maintain.

6.10 CONCLUSION

The use of IoT in geotechnical engineering is changing how we understand and monitor structural and environmental stability. Using IoT, traditional methods of data collection can now be substituted for real-time data and advanced sensor technologies that all contribute to the amplification of geotechnical data. By applying this transformation, engineers will be able to address long-standing geotechnical challenges with unprecedented precision and insight into structural health monitoring, landslide prediction, and subsurface analysis. Its parts are some of the crucial elements of the various IoT such as WSNs, automated data procurement systems that always improve big efficiencies but also can lead to cutting a high level of risk and increase security adequacy. AI, ML, and cloud-based analytics provide predictive insights as well as early warning systems which are vital for disaster prevention and infrastructure safety on IoT solutions. This combination of legacy geotechnical practices with smart IoT automation covers the entire base of engineering problems, combining the tested reliability from traditional methods and the accuracy and scalability of smart technologies. Technological advances in areas such as blockchain for data security; edge computing for real-time analysis; and smart systems that can monitor soil, slopes, and foundations in near-real-time will be central to the discipline moving forward. Spectrums such as improved decision-making, sustainability, and risk mitigation which we have in this day still unrealized by many could be tapped to many greater extents through IoT showcasing a resilient climate action in geotechnical perils years ahead.

REFERENCES

Abbasnejad, B., Soltani, S., Karamoozian, A., & Gu, N. (2024). A systematic literature review on the integration of Industry 4.0 technologies in sustainability improvement of transportation construction projects: State-of-the-art and future directions. *Smart and Sustainable Built Environment, 45*, 37.

Broekman, A., Jacobsz, S. W., Louw, H., Kearsley, E., Gaspar, T., & Burke, T. S. D. S. (2020). Fly-by-Pi: Open source closed-loop control for geotechnical centrifuge testing applications. *HardwareX, 8*, e00151.

Chawla, R., Singhal, P., & Garg, A. K. (2020). Internet of things driven framework for smart solar energy system. *Journal of Energy Resources Technology, 142*(1), 011201.

Dwivedi, D., Lepková, K., & Becker, T. (2017). Carbon steel corrosion: A review of key surface properties and characterization methods. *RSC Advances, 7*(8), 4580–4610.

Firoozi, A. A., Tshambane, M., Firoozi, A. A., & Sheikh, S. M. (2024). Strategic load management: Enhancing eco-efficiency in mining operations through automated technologies. *Results in Engineering, 24*, 102890.

Hong, S. C., Lee, W. G., Kim, W. J., Kim, J. H., & Jung, J. P. (2011). Reduction of defects in TSV filled with Cu by high-speed 3-step PPR for 3D Si chip stacking. *Microelectronics Reliability, 51*(12), 2228–2235.

Jacobsz, S. W., Standing, J. R., Mair, R. J., Hagiwara, T., & Sugiyama, T. (2004). Centrifuge modelling of tunnelling near driven piles. *Soils and Foundations, 44*(1), 49–56.

Kulkarni, M. B., Ayachit, N. H., & Aminabhavi, T. M. (2022). Biosensors and microfluidic biosensors: From fabrication to application. *Biosensors, 12*(7), 543.

Pal, A., Kulkarni, M. B., Gupta, H., Ponnalagu, R. N., Dubey, S. K., & Goel, S. (2021). Portable and autonomous device for real-time colorimetric detection: Validation for phosphorous and nitrite detection. *Sensors and Actuators A: Physical, 330*, 112896.

Rane, N., Choudhary, S., & Rane, J. (2024). Contribution of ChatGPT and Similar Generative Artificial Intelligence in Geotechnical Engineering and Soil Mechanics. *Available at SSRN 4681735*.

Vaish, R., Dwivedi, U. D., Tewari, S., & Tripathi, S. M. (2021). Machine learning applications in power system fault diagnosis: Research advancements and perspectives. *Engineering Applications of Artificial Intelligence, 106*, 104504.

Villa, V., Naticchia, B., Bruno, G., Aliev, K., Piantanida, P., & Antonelli, D. (2021). Iot open-source architecture for the maintenance of building facilities. *Applied Sciences, 11*(12), 5374.

7 Internet of Things-Based Slope Examining in Geotechnical Engineering

7.1 INTRODUCTION TO IoT IN SLOPE STABILITY ANALYSIS

Slope stability is a critical factor in geotechnical engineering, particularly in the context of mining operations. Slope failures in open-cast mines can have disastrous consequences, not only in terms of operational delays and financial losses but also in posing significant risks to human life and the environment.

7.1.1 Fundamentals of IoT for Slope Monitoring

Slope monitoring IoT systems are majorly a network of IoT devices coupled with sensors and communication technologies that acquire data from the field in real-time (Table 7.1).

7.1.2 IoT Systems and Components

An IoT system for geotechnical monitoring, in essence, consists of sensors buried into the slope and real-time wireless data collection units and communication modules. These devices are connected to cloud or on-premise platforms to carry out a continuous

DOI: 10.1201/9781003537625-7

TABLE 7.1 Fundamentals of IoT for slope monitoring

FIELD	*DESCRIPTION*	*KEY COMPONENTS*	*APPLICATIONS*	*CHALLENGES*	*TECHNOLOGICAL SOLUTIONS*
Sensors and data acquisition	Sensors play a crucial role in monitoring slope stability by collecting real-time data on factors like soil movement, moisture, temperature, and vibration	Accelerometers, inclinometers, GNSS, soil moisture sensors	Monitoring of landslides, erosion control, infrastructure safety	High installation costs, sensor maintenance in harsh environmental conditions	Low-power wireless sensors, rugged sensors with long battery life, real-time data transmission
Wireless communication protocols	Essential for transmitting sensor data to the central server or cloud system in real-time, enabling timely monitoring and alerts	LoRaWAN, Zigbee, NB-IoT, cellular communication	Data collection from remote, hard-to-access slopes	Interference from environmental factors, limited coverage in remote areas	Multi-hop networks, use of low-power wide-area network (LPWAN) technologies to cover large monitoring areas
Data processing and analytics	Raw data from sensors must be processed and analysed using IoT platforms to derive meaningful insights, predict slope failures, and issue alerts when thresholds are met	Edge computing, AI/ML algorithms, predictive analytics	Early warnings for landslide events, preventing disasters, infrastructure planning	Handling large volumes of data, false alarms due to noise in data	AI-based filtering and anomaly detection, cloud computing for big data analysis, predictive models
Power management	Ensuring long battery life for sensors in remote areas is critical for continuous slope monitoring, as sensor maintenance or replacement can be challenging	Solar panels, energy harvesting, low-power sensors	Sustainable monitoring in energy-limited environments	Limited solar exposure, battery drainage in harsh weather conditions	Energy-efficient designs, energy harvesting systems, dynamic power management techniques
Data visualization and decision support	Data visualization tools help in understanding the real-time slope conditions and making informed decisions based on historical and current data trends	Dashboard interfaces, GIS systems, data loggers	Enabling decision-makers to assess risk levels, manage slope safety programs	Interpreting complex data into actionable insights, real-time data integration	Cloud-based dashboards, integration of GIS and real-time analytics, AI-assisted decision-making platforms

analysis of data and graphical representations. They combine to create a holistic and automated system, allowing the engineer to remotely control slope stability behaviour, alert in case of deviation, and anticipate threats. Core sensors in slope monitoring include (a) tiltmeters, (b) inclinometers, and (c) strain gauges (Smethurst et al., 2017).

The backbone of IoT-based slope monitoring is sensors. Tiltmeters, inclinometers, and strain gauges are some of the most commonly used devices. When a slope begins to move, it causes an angular soil tilt that can be recorded by tiltmeters. Inclinometers, however, measure lateral movements in the slope and are typically buried deep in order to monitor subsurface displacements. Strain gauges measure the deformation of a material, which is vital for predicting the areas where stress exists and can cause slope failure. Altogether, these sensors afford a broad view of both surface and subsurface data for comprehensive slope stability studies (Yadava et al., 2020).

7.1.3 Importance of Slope Stability in Geotechnical Engineering

It is essential to know how slopes behave and predict if they might collapse in mining (and many other industries). Typically, slope stability monitoring is planned at regular intervals; however, this approach leads to late detection of the changes, and in most cases, it just misses real-time variations happening inside the slopes. Given this, continuous monitoring systems are mandatory in any modern geotechnical practice as mining operations develop on a larger scale and extend to greater depths (Akbulut et al., 2024).

7.1.4 The Role of IoT in Improving Conventional Slope Monitoring Methods

Competent slope stability remains a vital factor for construction and urban planning. Continuous and real-time data from different sensors inside the mine's geological structures are collected through IoT (Hou, 2018). This data is also used to monitor subtle shifts in soil movement, changes in moisture levels, pressure points, etc. IoT technology offers a more immediate, accurate perspective that is integrated with existing methods that tend to work at a glacial pace, not proactively or fast enough when we rely only on human capital. IoT enables engineers to not just monitor slopes better but also stay prepared for emerging risks and address them in a timely manner to prevent slope failures from turning catastrophic (Abraham et al., 2020).

7.1.5 Benefits and Barriers of IoT Slope Monitoring

Although IoT-based monitoring has some inherent superiority, such as prompt data gathering and better prediction power, it tends to bring diverse challenges. One of the key features that IoT provides is the ability to capture vast volumes of data from different points along a slope, thus offering a holistic measure of its stability. This allows for more accurate predictions and quicker interventions. That said, operational challenges can threaten the wider implementation of this technology such as maintaining network coverage in remote mining areas, real-time data management and processing at scale, and the significant upfront costs associated with installing an IoT infrastructure (Dewoolkar et al., 2003; Romdhane et al., 2017). However, given the long-term gains of meeting slope failure risks head-on and lowering operational costs in a more efficient manner, IoT remains an invaluable aid for geotechnical engineering.

7.2 WIRELESS COMMUNICATION IN IoT-BASED SLOPE MONITORING

IoT systems transmit data wirelessly, allowing for monitoring slopes in real-time without having anyone go for manual data retrieval. Sensors can communicate their data to a central hub or directly to cloud-based systems using wireless communication technologies such as LoRaWAN, Wi-Fi, and cellular networks. The decision on the kind of communication technology to be employed will vary with factors such as distance from the monitoring site, terrain features, and the extent of the existing network. Wireless communication needs to provide a way to monitor the conditions in remote or hazardous areas that cannot support traditional wired systems.

7.2.1 Linking IoT with Traditional Geotechnical Approaches

IoT is not a substitute for conventional geotechnical methods; it rather functions to supplement the same with an extra layer of data and knowledge. For example, combining IoT data with traditional slope stability models enables engineers to advance their interpretation by embedding real-time slopes into prognostic models. IoT rests on top of the existing traditional surveying and monitoring methodologies. It can be called a supplementary approach enhancing those practices by providing continuous monitoring over slopes, reducing the need for manual surveys, and creating a density/pattern of sensors over slopes, filling up gaps between inflation intervals. This would require frequent manual observation/replacement, which would make slope failures more predictable and less likely a surprise. The integration strengthens the safety and efficiency of mining operations by enhancing personnel tracking, vehicle data transfer in automated mines, energy system optimization, and mine estimating (Mustafa et al., 2013).

7.2.2 Slope Stability Risk Factors and IoT Monitoring Solutions

Open-cast mine slope failures are governed by a variety of natural and induced risk factors. Crudely speaking, natural factors such as rainfall, seismic activity groundwater fluctuation, or erosion weaken the strength of slopes over time.

$$R_s(t)=\sum_{i=1}^{N}\left(\alpha_i\cdot\frac{S_i(t)}{W_i(t)}+\beta_i\cdot\frac{P_i(t)}{T_i(t)}+\gamma_i\cdot\frac{D_i(t)}{G_i(t)}\right)+\eta\cdot\frac{1}{\log(1+F(t))}$$

$$S_i(t)=\frac{\partial\sigma_i(t)}{\partial t}+\lambda\cdot\frac{\partial^2\sigma_i(t)}{\partial t^2}$$

$$W_i(t)=\frac{\int_0^T\frac{\partial\omega_i(t)}{\partial t}\cdot\left(1+\log\left(1+\frac{\partial\omega_i(t)}{\partial t}\right)\right)dt}{\int_0^T\omega_i(t)dt}$$

$$P_i(t)=\frac{\partial\pi_i(t)}{\partial t}+\mu\cdot\frac{\partial^2\pi_i(t)}{\partial t^2}$$

$$T_i(t)=\frac{\log\left(1+\int_0^T\frac{\partial\theta_i(t)}{\partial t}dt\right)}{\int_0^T\theta_i(t)dt}$$

$$D_i(t)=\frac{\partial\delta_i(t)}{\partial t}+\kappa\cdot\frac{\partial^2\delta_i(t)}{\partial t^2}$$

$$G_i(t)=\frac{\int_0^T\delta_i(t)\cdot\frac{\partial\delta_i(t)}{\partial t}dt}{\log\left(1+\frac{\partial\delta_i(t)}{\partial t}\right)}$$

$$F(t)=\sum_{j=1}^{M}\frac{R_j(t)}{C_j(t)}+v\cdot\frac{d^2R_j(t)}{dt^2}$$

$$C_j(t)=\frac{\int_0^T\chi_j(t)\cdot\frac{\partial\chi_j(t)}{\partial t}dt}{\log\left(1+\frac{\partial\chi_j(t)}{\partial t}\right)}$$

Symbols Explanation:

- $R_s(t)$ is the slope stability risk score at time t.
- $S_i(t)$ is the stress contribution from sensor i at time t.

- $W_i(t)$ is the water saturation factor at sensor location i.
- $P_i(t)$ is the pressure reading from sensor i.
- $T_i(t)$ is the tilt angle or temperature measurement from sensor i.
- $D_i(t)$ is the displacement at sensor location i.
- $G_i(t)$ is the ground shear stress at sensor location i.
- $F(t)$ is the combined external force acting on the slope.
- $C_j(t)$ is the counteracting force from support systems at location j.
- $\sigma_i(t), \omega_i(t), \pi_i(t), \theta_i(t), \delta_i(t), \chi_j(t)$ are stress, water, pressure, tilt, displacement, and counteracting parameters measured by the sensors at time t.
- $\alpha_i, \beta_i, \gamma_i$ are weighting coefficients for different sensor contributions.
- λ, μ, κ are regularization parameters for stress, pressure, and displacement factors.
- η, v are regularization parameters for force terms.
- N, M are number of sensors and a number of support systems.

7.3 IoT ADOPTION IN EARLY-STAGE LANDSLIDE AND SLOPE WARNINGS

Early warnings in case of slope failures and landslides are being drastically transformed through IoT tech. As IoT plays a significant role in land monitoring, it implements a network of sensors along the slope surface that helps to monitor soil movement, stress, and water content from time to time. If these parameters are defined within predefined thresholds, the system can in turn automatically trigger alerts to let engineers and decision-makers know about potential slope instability. These proactive steps ensure that timely interventions like slope stabilization or evacuation take place, which further reduces the possibility of accidents and operational failures.

7.3.1 Monitor Rainfall, Groundwater, and Soil Movement in Near-Real-Time

In the realm of monitoring, there is no doubt that IoT and edge computing can contribute to valuable insights into environmental conditions influencing slope stability. A rainfall sensor helps to monitor the amount of precipitation, identical to a soil moisture sensor, which supports in soil saturation, leading to landslides. The sensors for groundwater keep tabs on changes in water levels, which may add pressure to the structural scale of the slope itself, whereas soil movement sensors are designed to measure the mass movements of the slope. When used to analyse these data points in real-time, IoT systems form a holistic picture of the risk landscape, as it changes and fluctuates in order to provide more precise readings used for assessments and responses.

7.3.2 Slope Stability Using IoT Data with the Help of Predictive Analytics

Not only does the data created by IoT systems help predict and alert potential slope instability, but these predictive analytics can further be used for various risk assessment practices. By leveraging machine learning, algorithms can process historical and real-time data coming from IoT sensors to detect patterns and correlations between slope failure and different risk factors. The predictive nature of these models allows for predicting the likelihood of slope failures as conditions change and serves to aid in risk management. Engineers can use IoT data and predictive analytics to make informed decisions on steps they need to take to avoid slope instability and allow mining operations cost-effectively and safely.

7.4 SLOPE MOVEMENT MONITORING USING SENSORS

Inclinometers for sensing subsurface lateral movements, tiltmeters for assessing surface angular shifts, and extensometers for tracking deformation are popular types of sensors (Huang et al., 2021). In some cases, rain gauges, piezometers (for measuring the groundwater level), and vibrating wire strain gauges are also deployed on a site to monitor environmental influences on slope instability. Including data from different methods means that each sensor measures something different, but in total, they give a fairly complete monitor of the entire area and its smallest slippage.

7.4.1 Field Deployment and Calibration of IoT Sensors

Using IoT sensors in a mining and geotechnical field, of course, needs good planning for taking the data accurately. Sensors must be deployed where slope movement or instability is likely to happen, e.g., at the crest, base, or middle of a slope. For reliable measurements, it is crucial that sensors be calibrated correctly before they are deployed. This step involves recalibrating the sensors to the actual site conditions and verifying if they can respond accurately to changes in slope behaviour (Ma & Guo, 2020). In severe mining environments where conditions can change rapidly, periodic recalibration may be required to continue keep exactitude.

7.4.2 Data Acquisition on Remote Sites: Tools and Techniques

After these IoT sensors are deployed, they must collect data constantly, and this data needs to be sent to a master system for further analysis. Remote data acquisition, meanwhile, is achieved through methods such as wireless mesh networks, satellite communication, or cellular networks, depending on the site location and infrastructure (Guilhot et al., 2021). It is generally real-time data that is collected and available through cloud-based platforms for visualization by engineers. They can analyse this data remotely from anywhere, detect trends, and respond to events. Tools allow remote monitoring, which is easier than ever before because now it is at your place where you could not be physically at the site.

While IoT sensors have not been utilized to monitor slope stability in an operational mine yet, there are several case studies that demonstrate the effectiveness that these devices could provide and the extent to which they could reduce the risks in mining operations. One such slope monitoring application included the installation of a network of tiltmeters and piezometers in an open-pit mine in South America to continuously determine the stability of a full bench-scale unstable slope. Thanks to the real-time data collected by these sensors, engineers could predict a slope failure and get everybody out of the way in time, meaning there was no loss of life nor any equipment ruined. In another instance, a mining site in Asia implemented IoT-based monitoring systems to keep up with rainfall and soil moisture levels at a remote mine to mitigate landslides using this data to stabilize slopes prior to high rainfall events (Lau et al., 2023). In the case studies, we find the IoT sensors to be a life-saving technology because they are very reliable in their geotechnical operations.

7.5 AMPLIFICATION IN IoT-BASED SLOPE MONITORING

Real-time data collection is key to making quick decisions to reduce risk while doing IoT-based slope monitoring. These field data sensors, such as inclinometers, tiltmeters, piezometers, and flat-outs, regularly collect data on slope movement and strain in soil or rock, as well as fluctuations of water levels in the ground and environmental metrics as shown in Figure 7.1.

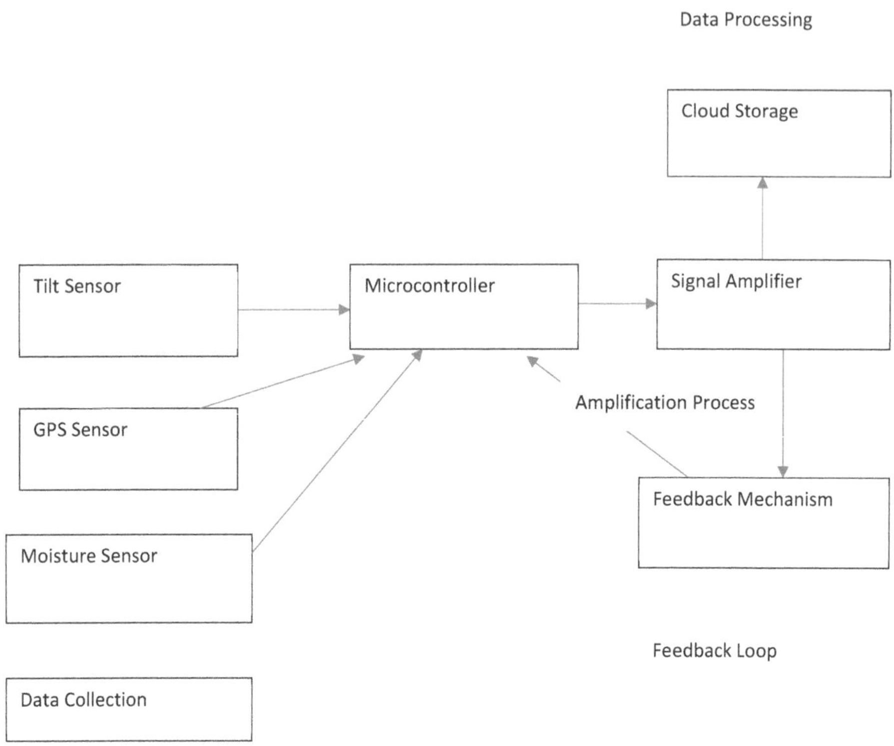

FIGURE 7.1 Amplification in IoT-based slope monitoring.

7.5.1 Slope Monitoring Sensors: Real-Time Data Capture

This data is then sent to a central monitoring system immediately via wireless communication networks. By capturing in real-time, engineers are able to monitor slopes continuously and take immediate corrective action when the first signs of instability appear guaranteeing better risk prevention in mining environments.

7.5.2 Data Augmentation Techniques to Improve Accuracy and Precision

Amplification techniques are often used to build data resolution with higher accuracy and precision. Data amplification: To drift raw sensor data, remove noise, and correct measure errors, the produced information has to be reliable. For example, signal amplification techniques are employed to intensify low-level signals at distant sensors, and more sophisticated routines can combine data collected from multiple instruments in order for improved slope safety templates. Margin checking and correction detection are paramount in the mining environment, where factors such as interference from electromagnetic sources and unreliable GPS signals due to terrain or weather conditions are likely to cause sensor errors.

7.5.3 Slope Stability Storage and Analysis with Cloud-Based Solutions

For the management and storage of these huge amounts of data that IoT sensors create, cloud-based platforms are the king. Slope monitoring sensors' cloud technology is then fed into centralized databases in the easiest possible way to serve telecommunications systems, so that real-time data access and transfer becomes a piece of cake. By enabling the storage of large amounts of historical data for long-term analysis, trend detection can leverage the scalability that cloud platforms provide. They also provide a state-of-the-art analytical tool for slope stability data processing where any engineer can perform live assessments, visualize trends, and predict potential failures in the future using machine learning models.

7.5.4 Edge Computing for On-Site Processing in Slope Monitoring

On one hand, the storage and analysis-related engine from cloud solutions allow us to use a robust and scalable solution for all data management. On the other hand, edge computing provides an important value in the data processing part. Edge devices (i.e., local servers or gateways) perform the processing for IoT-based slope monitoring at or near sensor locations, thus limiting the need to always send sensor data to the cloud. This eliminates latency, ensuring that they can quickly respond to urgent slope condition changes across their network. This enables edge computing systems to automatically sound real-time alarms and take preventive actions based on sensor data being collected rather than dependent on the latency of sending data back into a cloud for analysis. For high-risk mining operations like these, the nearly immediate response can help avert slop failures, saving lives in the process.

7.6 ADVANCED ANALYTICS FOR SLOPE MONITORING USING IoT

In IoT, field systems continuously monitor the slope and stream observations to a data zone where they are stored as mobile landscape large datasets (see Image 3) requiring advanced analytics for interpretation; analytics are often required from cloud services.

Big data analytics assist in deriving valuable insights from these humongous datasets (Wang et al., 2014).

7.6.1 IoT-Based Slope Monitoring Using Big Data Analytics

Patterns and trends in slope stability can be detected through data mining and statistical analysis to facilitate planning in advance. Big data analytics has a payout in long-term slope monitoring; with historical data that can be compared to current inputs, it predicts slope behaviour, which eventually contributes to an improvement in risk assessment of mining operations (Nandy et al., 2023).

7.6.2 Machine Learning Algorithms for Slope Failure Prediction

Slope monitoring with machine learning (ML) as advanced analytics model input data is sensor-based, and model output data is the prediction of slope failures using various ML algorithms such as regression models, decision trees, and neural networks. These algorithms process historical and real-time data from IoT devices, discovering patterns that may cause slope instability. For example, one of the ways artificial intelligence could be used is for models based on ML that can correlate factors, say, rainfall intensity and saturation levels with ground deformation to forecast a landslide. Mining engineers can use ML for predictive analytics to prevent catastrophic failures (Phoon et al., 2019).

7.6.3 Visualization of IoT-Based Slope Stability Data

It is important to visualize slope stability data properly so that the complex data you will be dealing with can be translated into actionable knowledge. Data visualization tools ensure that real-time sensor data is displayed in an easy-to-understand format, such as graphs, heat maps, or 3D models for IoT-based monitoring systems. It provides a visual representation to the engineers, helping them quickly see how the slope currently looks. This process makes it easy and efficient to point out those slopes that engineers may further need to check. Using visualization of the data, stakeholders are better able to understand trends, potential hazards, and the effectiveness of stabilization measures that have been implemented, allowing them to communicate and make informed decisions.

7.6.4 Techniques of Filtering, Cleaning, and Normalizing Data

Raw sensor data has to be filtered, cleaned, and normalized before we can perform advanced analytics on it. It often employs real-time filtering and cleaning techniques to filter noise and irrelevant information from the data stream and clean out misbehaving sensors that inject outliers into the data. Normalization—standardizing the data from multiple sensors to a common scale—is executed to facilitate easier processing. Pre-processing plays a critical role in making the dataset well-conditioned, thereby determining the reliability of results related to slope stability analysis and prediction. These types of techniques not only help read the data more accurately, but they also improve both real-time monitoring and longer-term predictive modelling by providing a controlled quality level of data.

7.7 AUTOMATION OF IoT-BASED SLOPE MONITORING SYSTEMS

One of the main practice-impacting features of IoT slope monitoring is that it alerts and notifies staff automatically when movements exceed predetermined levels.

7.7.1 Slope Movement Detection Automated Alerts and Notifications

This information can alert engineers and decision-makers to early warning signs of instability with ground movement or water levels on the rise, for example, in real-time via email, SMS, and specialized monitoring dashboards. These early warnings provide true alert functionality so that mining operations can react and intervene before catastrophic slope failures take place.

7.7.2 IoT and Remote Monitoring Control Systems

IoT-based systems provide complete remote operation, which means that engineers can now monitor slope stability from anywhere. The cloud platforms enable real-time data from the sensors and allow for detailed reports and visualizations on slope conditions without the requirement of human presence on-site. Furthermore, remote systems mean that such corrective measures can now also be remotely controlled as necessary, for example, modifications to drainage systems or activation of stabilization mechanisms—improving operational efficiency and safety, especially in mines found in hazardous or hard-to-reach areas.

7.7.3 Integration of Automated Surveying with Geotechnical Monitoring

The surveying process allows for automation that enables better accuracy and data co lection frequently when carrying out a slope stability analysis. These automated su veying technologies are able to be integrated with IoT-based geotechnical monitorin systems, continuously conducting surveys and mapping of slope changes. Such survey generate accurate 3D ground models, which, when coupled with sensor informatio create a complete insight into slope stability. Automated monitoring linked with Io technology guarantees that all minor slope movements are assessed instantly to ale more informed decisions.

Case Studies: Automation in IoT-based slope monitoring systems has demonstrate the benefits of automation in slope monitoring in real-world projects. In a mining pro ect in Australia, a high-risk slope monitoring system was implemented based on Io and with automated alerts and drone-based surveying. This system recorded the groun movement early enough, allowing for timely intervention to prevent a potential landslid In another European project, automated water drainage systems that used real-time ser sor data were integrated with the existing system and were very effective in comba ing groundwater-induced slope failures. These case studies demonstrate that automatio in slope monitoring based on IoT leads to safer and more reliable operations in gec technically sensitive environments. IoT-driven safety and risk management for slope ensures constant monitoring of factors inducing slope stability, such as soil movemen water pressure, and seismic activity. The sensors collect real-time data, which is prc cessed and analysed continuously to develop dynamic maps that show areas with hig risk. Real-time risk assessment and hazard mapping help engineers and safety staff t understand the rapidly changing situations and identify the areas that require immediat action. Real-time risk assessment also leads to proactive decision-making, reducing th chances of unexpected slope failure and thus enhancing the site's general safety.

7.8 DEVELOPING IoT-BASED EARLY WARNING SYSTEMS

In high-risk mining areas, IoT-based early warning systems are important to avoi disasters. Such systems merge real-time sensor data with predictive analytics to war imminent slope failures. For instance, if sensors record abnormal behaviour in groun movements or groundwater levels, the system can automatically raise an alarm an notify stakeholders. Remote or undeveloped cast mines, in particular, may derive sig nificant benefits from their use, as traditional methods of monitoring can be incomplet and such systems serve as an additional secure line of defence against the occurrence o a sudden geotechnical hazard.

7.8.1 IoT-Based Emergency Response Plans for Slope Failures

When a slope fails, IoT systems could enhance emergency response capabilities substantially. These systems broadcast relevant information about the state of the slope condition, thus allowing for timely decisions and responses. IoT systems, on the other hand, can monitor where debris or mudslides are moving and relay that information to emergency crews, showing them the fastest way out or areas to avoid in a wildfire. Furthermore, automated systems can automatically initiate mitigative measures (e.g., prevent machines and tools from operating or locking hazardous chemicals) to limit the risks associated with slope failures on human health and equipment.

7.8.2 Risk Management Strategies Using IoT in Slope Monitoring

Numerous case studies prove the suitability of IoT risk management in slope monitoring. An IoT solution was established to track rainfall, soil moisture, and ground displacement among mountains susceptible to landslides. The information from the monitoring data enabled to establish a real-time hazard map and thus evacuate the workers in due time, avoiding fatalities. At another mine in South America, a large open-pit showpiece operation deployed early warning systems using IoT and emergency response protocols to anticipate slope failures and set new safety benchmarks. These examples illustrate how IoT plays a part in improving safety and risk management within the geotechnical and civil construction arena.

7.9 IoT-BASED GEOTECHNICAL SLOPE MONITORING PROJECTS

Slope monitoring in infrastructure projects with IoT real-time monitoring of slopes in large-scale construction projects, such as highway cuttings, tunnels, and high-rise buildings, ensures both stability and safety.

7.9.1 Non-Performance in Slope Monitoring for Infrastructure Projects

For example, IoT sensors placed in and around the excavation sites can monitor soil movement, groundwater levels, and environmental conditions 24/7. This data makes it possible for engineers to anticipate dangers and use countermeasures before landslides or structural failures occur. This proactive monitoring can aid in avoiding project slip, being that safety does not have to wait until the end of a phase.

7.9.2 IoT in Monitoring Slopes in Mining and Quarrying Operations

For instance, in the mining and quarrying industries, IoT-based slope monitoring systems are used to monitor the stability of open-pit mine and construction walls. If the sensors that are placed on slopes observe changes in ground movement, pressure, and a few other critical factors, then this real-time data will be available to avoid any catastrophic failure. In an open-pit mining operation, for example, IoT systems can monitor rock stability and vibrations and alert operators of any early signs of destabilization before things get worse. This is valuable when it comes to ensuring mining operations are optimized for maximum output while also keeping workers safe and reducing the environmental impact.

7.9.3 IoT for Dam and Embankment Slope Stability

Sensor technologies, such as those associated with IoT, are used to monitor the stability of slopes, dams, and embankments, protecting against catastrophic flooding when sloped ground fails. By accurately monitoring features including pore water pressure, soil displacement, and structural strain in real-time, IoT sensors can be deployed on any site. In a dam monitoring project, for example, IoT systems detect early signs of trouble by closely following the movements in the foundation and body of the dam. This makes the necessary maintenance and reinforcement measures possible in time so that these critical components of infrastructure are safe and usable. Powering IoT devices in the slope monitoring locations, often in a remote or difficult topography, can be a daunting task. With traditional power sources out of the question, or highly impractical and even impossible, solutions need to be avant-garde.

7.10 REMOTE SLOPE MONITORING DEVICES POWERED BY IoT

Some of the widely used methods include using solar panels and wind turbines to power IoT devices with renewable energy. The system will run without stopping in this case, with which you must design a battery-operated system that needs heavy-duty batteries or run on an alternate power source. There can be automatic power-saving systems, thus reducing the chances that sensor replacement or maintenance (especially in Wireless sensor network (WSNs) powered by batteries) occur frequently.

7.10.1 Slope Monitoring Energy Harvesting Methods

Energy harvesting techniques provide creative solutions for IoT sensors to be supplied with power in slope monitoring applications. This form of energy harvesting utilizes things like the movement of a slope or machinery to create electrical power through methods such as vibration energy harvesting. Thermoelectric generators (e.g., see KIDS-D01 and e-Panel Final) convert temperature differences into electrical energy, piezoelectric devices, etc. These approaches can help sensors to operate longer without human intervention or recharge, allowing them to be deployed in remote environments where replacing batteries as they deplete would not be practical.

7.10.2 Low-Energy IoT Networks: Long-Range Communication Solutions

Efficient communication between sensors is challenging, and long-range communication solutions are key in IoT-based slope monitoring systems especially when power resources are scarce. Solutions such as low-power wide-area networks (LPWA) like LoRaWAN or SigFox offer low-energy long-range IoT connectivity. Our solutions allow significant distances to be covered with minimum energy consumption and are perfect for monitoring over a greater scale or across wider slope types of areas. Mesh networks also improve connectivity and reliability, as they enable sensors to pass their data through one or more nodes (forwarders), thus extending the coverage area and guaranteeing robust communication.

7.10.3 Geotechnical Monitoring over IoT Devices: Battery Life Optimization

In deep geotechnical monitoring of IoT devices, maintaining operational efficiency and reliability requires optimizing battery life. Two essential components in prolonging battery life are energy-efficient sensors and low-power communication protocols. Power consumption can be minimized to a greater extent by methods like duty-cycling (alternating between active and sleep modes with respect to the sampling interval) and adaptive sampling (data is sensed/written according to the surrounding context). Regular battery health checks and the use of high-capacity batteries designed for extended life can also minimize monitoring system downtime, requiring less maintenance.

7.10.4 Cybersecurity Challenges in IoT-Based Slope Monitoring System

Many things that contribute to the insecurity come from IoT sensors all over being installed and transmitting data 24/7. Securing these environments depends on strong encryption, secure data authentication methods, and frequent security updates to counteract new threats. Moreover, it is very important to have security built into the system design from day 1 to minimize risks.

7.11 BLOCKCHAIN AND OTHER TECHNOLOGICAL ASPECTS FOR IoT SLOPE MONITORING SECURITY

Indeed, blockchain technology presents a potential remedy to secure and ensure the integrity of IoT slope monitoring systems. On the one hand, blockchain offers a decentralized and immutable ledger; details captured by IoT sensors are recorded in a tamper-proof and transparent manner. This is where each entry of data is connected to the previous one in a chain, making it secure against tampering. Additionally, secure multi-party computation and homomorphic encryption are other technologies which can be used to improve the security of privacy-related data. Existing technologies, which your budget will probably cover, can allow the processing of sensitive data and its analysis without it being seen by unauthorized individuals, consequently increasing the security of monitoring systems.

7.11.1 Best Practice in the Deployment of IoT-Based Ski Monitoring Systems

For those deploying IoT-based systems for slope monitoring, it is crucial to follow best practices to ensure the solutions are effective and reliable.. Central practices comprise proper site assessments determining the preeminent sensor types and deployment locations. Sensors are maintained and calibrated regularly to collect accurate data. System integration testing should also be in place to ensure that all the parts will work together. Providing training on monitoring systems and response to emergencies helps to fully improve the deployment as well so that the system may work efficiently.

7.11.2 Protocols for Data Handling and Security in IoT Applications

The integrity and confidentiality of data collected through IoT applications depend on strong protocols related to both data management and security. To ensure data is correctly and readily recorded, your data management protocols should define how you collect, store, and analyse it. It includes all those best security practices from encryption to secure data transmission or compliance with other standards and aspects; it also ensures that the access does not become accessible and emerges into unauthorized availability, hence providing data protection against unauthorized access. You should also know your backup and recovery plans to remain prepared for the exact nature of systems failure or cyber-attack.

7.12 CASE STUDIES OF SLOPE STABILITY MONITORING: IoT-BASED

In mountainous areas, such as mine sites, it is essential to ensure an optimal work environment, and IoT can easily be a solution for slope monitoring systems, which will be beneficial with respect to the unique challenges posed by steep and unstable terrain. For example, a case study is put forward about the installation of an array of IoT sensors in a mountainous landslide-prone region. This included placing tiltmeters, inclinometers, and strain gauges to monitor both ground movement and environmental conditions. Slope monitoring to provide early warning signals and preventive measures were made easy as the data provided have fast access. Interactive System Ospreys code also integrated data from IoT devices that helped adjust the monitoring system for weather conditions, thus enhancing the accuracy of the slope stability forecasts to make people who live downstream much safer.

7.12.1 VR-Based IoT Application for Slope Monitoring in Roadways and Highways with IoT

For example, in roadways and highways, IoT has been successfully used to explore slope monitoring where it is utterly important for safe transportation that the slopes be stable. A highway case study, prone to soil erosion and landslides, used IoT sensors to periodically monitor slope conditions and environmental factors. The assessed data could support predictive (pre-fracturing) models of slope failures and timely intervention needing maintenance or reinforcement. Real-time monitoring was implemented, making it possible to respond to any problems detected almost instantly which greatly decreased the risks of road closures and increased safety on the route.

7.12.2 Monitoring Slops in Coastal and Riverbank Areas

There are several kinds of problems, like coastal zones and riverbank zones, whic have specific complications to solve, such as erosion and soil movement, making the particularly relevant due to the influence of water bodies on stability, sediment tran port, and land degradationThe use of IoT sensors to keep track of slopes in these sens tive regions is an example study case. Processes, such as soil moisture, groundwate or coastal erosion, have been investigated using deployed sensors. With this real-tim data, the Dielectric Absorption Ratio (DAR) had facilitated the development of erosio control mechanisms and slope failure early warning alarms. Utilizing weather forecas and tidal data, the monitoring system was able to provide a broad perspective of coast processes, which is vital for protecting coastal infrastructure and valuable ecosystems

7.12.3 Long-Term Slope Stability Monitoring by Leveraging IoT in the CAPEX-Heavy Projects

One of the main case studies could be monitoring the long-term slope stability in criti cal civil engineering projects, like dams and large-scale construction sites. The slop started to move due to the hydro saturation of the soil in the Setagaya Yukawa are Long-term movement trend data were obtained by introducing several IoT sensors fc monitoring purposes. The collected data were sent to be examined for trends and pat terns on the way it behaves to create measures for maintenance and risk. The case stud also showed that constant monitoring with the help of IoT can improve the safety an longevity of critical infrastructure by identifying problems in advance and allowing fo interventions to take place in a timely manner. This methodology not only enhance security but also optimized the planning of works and allocation of resources.

7.13 FUTURE TRENDS AND INNOVATIONS IN IoT FOR SLOPE MONITORING

The rapidly moving world of IoT-based geotechnical monitoring has introduced man new technologies, and these technologies are still under development. The developmen of sensor technologies, such as miniaturised sensors or nano-sensors, is becoming a emerging trend, where their higher sensitivity and precision can showcase the grea potential in monitoring the stability of slopes.

7.13.1 AI and ML Integrations for IoT in Slope Failure Prediction

IoT, with the help of artificial intelligence (AI) and ML, is changing the predictive slope failure analysis. AI/ML algorithms can sift through big data that is generated from IoT sensors and identify nuanced patterns and correlations between them, which could potentially be missed by more traditional methods, thereby creating value in the asset management process; for example, predictive analytics models can predict slope failures based on past data and real-time sensor inputs. Deep learning is one such method that looks particularly promising for improving the accuracy of predictions and adapting to changing circumstances. While further increasing the reliability of early warning systems, this integration also allows for increasingly dynamic management.

7.13.2 Next Step for IoT-Powered Slope Stability Research

There are a few other key aspects where future research in IoT-enabled slope stability might focus on. One such significant direction is to innovate advanced data analytics, such as leveraging quantum computing with the view of proficiently organizing humongous geotechnical data. The research will also investigate the integration of IoT with advanced modelling and simulation tools to more accurately predict slope performance under different conditions. Furthermore, the interoperability of IoT systems and geotechnical monitoring tools across different types will also be enhanced. They will also prioritize making data privacy and security stronger in order to counter new threats and safeguard confidential information.

7.14 CONCLUSION

These are just a few examples of why slope stability in spoil heaps on constructed cast mines is so important for mine operational safety and for preventing these catastrophic failures. Slope stability in mining is a highly complex problem because of the interplay between several geological and environmental factors but also as it has been generally addressed through manual surveillance and engineering appraisements. Nevertheless, these approaches are restricted and call for innovative technological alternatives. Combining the power of IoT and ML to monitor slope stability can offer a game-changing solution. With the additional live data capture capability of IoT devices, they are now able to use data+ML algorithms to predict potential areas for slope failures before they even happen. With the help of sensors, mining operations can be monitored 24/7, which provides them with a treasure trove of data that they can use to find trends and predict potential incidents. After looking at case studies and current use cases, we have surveyed successful implementations of IoT-based monitoring systems and ML models that can be used as important notes on the projects to be carried out. This may be more strident due to technical implications, e.g., poor-quality data, integration problems, etc., but the directions themselves are probably one of the very few hopeful areas in the real world. Healthcare AI solutions are slowly improving over time. In the future, developments in technology and new innovations are set to increase the potential of IoT and ML within the field of geotechnical engineering. Slope stability in mining has a future defined by improvements to these technologies, such as better data processing and application across different slopes so that safer yet more efficient practices are established on a global scale.

REFERENCES

Abraham, M. T., Satyam, N., Pradhan, B., & Alamri, A. M. (2020). IoT-based geotechnical monitoring of unstable slopes for landslide early warning in the Darjeeling Himalayas. *Sensors, 20*(9), 2611.

Akbulut, N. K., Anani, A., Brown, L. D., Wellman, E. C., & Adewuyi, S. O. (2024). Building a 3D digital twin for geotechnical monitoring at San Xavier mine. *Rock Mechanics and Rock Engineering, 9,* 1–18.

Dewoolkar, M. M., Goddery, T., & Znidarcic, D. (2003). Centrifuge modeling for undergraduate geotechnical engineering instruction. *Geotechnical Testing Journal, 26*(2), 201–209.

Guilhot, D., Hoyo, T. M. D., Bartoli, A., Ramakrishnan, P., Leemans, G., Houtepen, M., & Maknavicius, G. (2021). Internet-of-things-based geotechnical monitoring boosted by satellite instar data. *Remote Sensing, 13*(14), 2757.

Hou, X. (2018). Geotechnical engineering slope monitoring based on Internet of Things. *International Journal of Online Engineering, 14*(6), 165–176.

Huang, H., Liu, M., Mao, Y., Du, M., & Zhang, X. (2021). Research on road slope compound monitoring system based on internet of things technology. *IOP Conference Series: Earth and Environmental Science, 643*(1), 012161.

Lau, Y. M., Wang, K. L., Wang, Y. H., Yiu, W. H., Ooi, G. H., Tan, P. S.,... Chen, C. W. (2023). Monitoring of rainfall-induced landslides at Songmao and Lushan, Taiwan, using IoT and big data-based monitoring system. *Landslides, 20*(2), 271–296.

Ma, Y., & Guo, G. (2020). Intelligent algorithm of geotechnical test data based on Internet of Things. *Computer Communications, 158*, 32–38.

Mustafa, I. S., Din, N. M., Ismail, A., Omar, R. C., & Khalid, N. H. (2013). Antenna placement for landslide monitoring using analytical hierarchy process (AHP) and geographical information system (GIS). In *IEEE Symposium on Wireless Technology and Applications (ISWTA)* (pp. 295–300). Kuching, Malaysia: IEEE. https://doi.org/10.1109/ISWTA.2013.6688791.

Nandy, S., Adhikari, M., Ray, A., Rai, R., & Singh, T. N. (2023). Edge-centric intelligent early warning system for residual soil stability prediction in slope. *IEEE Internet of Things Journal, 11*, 3832–3839.

Phoon, K. K., Ching, J., & Wang, Y. (2019). Managing risk in geotechnical engineering–from data to digitalization. In *Proceedings of the 7th International Symposium on Geotechnical Safety and Risk (ISGSR)* (pp. 13–34). Singapore.

Romdhane, R. F., Lami, Y., Genon-Catalot, D., Fourty, N., Lagreze, A., Jongmans, D., & Baillet, L. (2017). Wireless sensors network for landslides prevention. In *IEEE International Conference on Computational Intelligence and Virtual Environments for Measurement Systems and Applications (CIVEMSA)* (pp. 222–227). India.

Smethurst, J. A., Smith, A., Uhlemann, S., Wooff, C., Chambers, J., Hughes, P., & Hughes, D. (2017). Current and future role of instrumentation and monitoring in the performance of transport infrastructure slopes. *Quarterly Journal of Engineering Geology and Hydrogeology, 50*, 271–286. https://doi.org/10.1144/qjegh2016-080.

Wang, H., Li, L., Jiao, Y. Y., Ge, X. R., & Li, S. C. (2014). A relationship-based and object-oriented software for monitoring management during geotechnical excavation. *Advances in Engineering Software, 71*, 34–45.

Yadava, D. K., Karthik, G., & Jayanthu, S. (2020). Studying time domain reflectometry to predict slope failure in open-cast mines. *Journal of Mining Science, 56*(5), 760–770. https://doi.org/10.1134/S1062739120057093.

8 IoT Application in Examining the Various Parameters of Soil and Ambient Constraints

8.1 IoT IN AGRICULTURE

We have seen a disruptive move when the Internet of Things (IoT) reaches out to farming as a transformation that easily combines how we manage our farms with providing operational and efficiency benefits. The deployment of interconnected sensors and devices for immediate monitoring and analysis of different agricultural parameters is possible through IoT technology. This ability to monitor and manage crops in precise ways makes many of the following use cases possible, which drive better crop yields and resource management.

Soil and ambient factors have great importance in the agriculture field. The quality of soil directly impacts growth and health, as well as nutrient availability, moisture levels, and fertility (Placidi et al., 2021). Temperature, humidity, and light also determine characteristically distinct ambient conditions that are vital in plant development and crop performance. Sustainably managing these factors is key to optimizing productivity and sustainability in agriculture (Dhanaraju et al., 2022).

The main goal is to investigate the application of IoT technology to monitor and alleviate soil as well as ambient limitations in agriculture. We will look at how some of the different soil types, which encompass composition, moisture level, temperature, and pH, play off against other outside influences such as temperature, humidity, and

DOI: 10.1201/9781003537625-8

ven light intensity. We have been examining IoT sensors and devices to help with eal-time soil and environmental monitoring and provide some context around where t is now versus when used well. In this chapter, we will explore a few of the basic IoT oncepts relevant to agriculture, as well as key technologies and their applications. We vill also address the pitfalls of urban soil monitoring and ambient monitoring through oT. Conclusively, the intent is to provide a holistic view of how IoT can boost precision griculture (PA) in overcoming limitations imposed by soil and environmental factors, ielping farmers operate more profitably and sustainably (AlZubi & Galyna, 2023).

Agricultural IoT is an application of the IoT, which involves collecting, transferring, and integrating data related to agricultural operations through a network of interconnected devices. This includes everything from soil moisture sensors and weather nonitoring devices to automated irrigation systems and smart machinery. IoT connects hese devices to a central network so farmers can access instant data about different ispects of their agricultural businesses in real-time; allowing better decision-making ind management than ever before (Chowhan & Dayya, 2019).

The potential of IoT in agriculture is not limited to data collection. This also nvolves advanced analytics and artificial intelligence integrated to process and analyse the data systematically, thereby enabling decision-making. This integration provides for fine-tuned management of irrigation mechanisms and the most efficient use of pesticides (~1%), in addition to other beneficial crop health and regulatory services. Similarly, in agriculture, IoT combines predictive analytics with historical data and real-time inputs to anticipate conditions and outcomes, leading to improved planning and risk mitigation (Kashyap & Kumar, 2021).

The main elements in IoT for agriculture consist of a number of sensors, different communication networks, and standalone data processing systems. Environment and soil condition monitoring systems, such as temperature, humidity, moisture, and nutrient content in soils, are very important factors that decide the appropriate growth of plants; therefore, it is necessary to monitor these environmental factors through sensors. These sensors are always gathering data and sending it across wireless networks to the central data hub (Castrignanò et al., 2020).

These data are transmitted through communication networks based on technologies like Wi-Fi, LoRaWAN (long-range wide-area network), or cellular networks from the sensors deployed in the field to cloud-based platforms where they can be leveraged. Upon the arrival of data, cloud computing and edge computing solutions hold these aggregated datasets for processing in order to deliver insights and heuristic responses under certain conditions based on predefined algorithms (Shaikh et al., 2022).

IoT in agriculture partners with technologies such as geographic information systems (GIS) for the analysis of spatial data, machine learning for predictive analytics, and automation to control irrigation systems and machinery. Such technologies reinforce one another to maximize the efficiency and accuracy of agricultural processes (Goel et al., 2021).

8.2 BENEFITS AND CHALLENGES

IoT in agriculture offers huge benefits. IoT data and analytics empower farmers wit real-time decision-making insights, resulting in higher crop yields, controlled use o resources, and lower operating costs. When it comes to irrigation, fertilization, or pesticide use, IoT enables PA, which results in the intelligent application of water, fertilizer and pesticides that is environmentally friendly.

Nonetheless, the deployment of IoT in agriculture is accompanied by a variety o obstacles. Concerns among firms about the hefty upfront investment for setting up Io networks in terms of sensors, connectivity, and data platforms are one of the mai hurdles. Not only this, but data security and privacy are paramount, as there can be ver sensitive agricultural data involved, and not everyone shall have access to it apart fro the threat of cyber threats.

The skills gap presents unique technical expertise required to manage and maintain IoT systems. IoT data may need to be turned into something a farmer or agricultur worker has the training needed to use and interpret. Additionally, environmental conditions such as severe weather and physical objects can contribute to decreased operational reliability due to the perspective of both sensor performance and inaccurate dat

While facing greater barriers in the uptake of IoT technology, there still remain a great potential to increase productivity, sustainability, and efficiency in farmin practices with the ongoing development of IoT technology given its applications i agriculture.

8.3 SOIL CHARACTERISTICS AND IMPLICATIONS

The fertility of the soil and its potential to support plant growth are real aspects o understanding soil health. This includes mineral particles, organic matter, water, an air. In view of chlorine's corrosive effect on pipes and the human respiratory system, d you really want a salt-free system? Water and nutrient holding capacities, aeration, roo penetration, etc., of the soil influence the health of crop plants and their yield (Akhte & Sofi, 2022).

Soil structure is the way in which soil particles (sand, silt, clay) clump together t form aggregates or clumps. Soil with a good structure has a mix of sand, silt, and cla that is stable in aggregates, promoting good soil aeration, drainage, and root growth Compacted soil stops natural aeration and water infiltration, which in turn can lesse root growth. To achieve higher crop yield, it is necessary that the farmer should hav knowledge of soil composition and architecture to use the right kind of farming practices and soil management strategies (Fuentes-Peñailillo et al., 2024).

8.3.1 Soil Moisture Content

The moisture content of the soil is one key factor that influences plant growth and hence farming productivity. The water content of soil—that is the amount of water that is available in the soil for plants to hydrate themselves and uptake nutrients—plays a crucial role. While moisture influences germination, root growth, and nutrient uptake, maintaining an optimal level of moisture is ideal rather than consistently causing underwatering or overwatering (Kalyani & Collier, 2021).

Too much water restricts oxygen to roots, which can also suffer from root diseases. On the other hand, a lack of moisture can incite drought stress, resulting in a loss in plant growth and yields. With the help of IoT technologies, such as soil moisture sensors, real-time information can be collected, enabling farmers to adjust irrigation practices more precisely based on the current status of water contents in their soils. This will help in good crop performance with optimal soil moisture balance, as well as water use efficiency, thereby saving valuable water resources (Kaswan et al., 2024).

8.3.2 Soil Temperature

There are two important factors that affect soil fertility and plant growth: the temperature of the soil and its pH. Seed germination, root growth, and microbial activity in the soil are all affected by soil temperature. The ideal temperature of soil varies by crop species, but it can be inhibitive at both ends for plant growth and yield.

Soil pH—the measure of soil acidity or alkalinity—influences the availability of nutrients as well as microbial life in various ways. The ideal pH range for many crops is 6.0 to 7.0. In the long term, strongly acidic or alkaline soils can result in nutrient deficiencies or toxicities and consequently affect both plant health and productivity. Through IoT sensors, soil temperature and pH are tested regularly, making the adjustments necessary, such as adding lime or sulphur to rectify pH imbalances required for optimal growing conditions (Chithaluru et al., 2023).

8.3.3 Soil Nutrient Levels

To sustain plant growth and the production of high crop yield, proper soil nutrient levels are important. Plants need various nutrients in different amounts—elements like nitrogen, phosphorus, and potassium. These nutrients, when occurring in excess or deficiency, can adversely affect the growth of plants, hence impacting yields and making the crops susceptible to pests and diseases (Karunathilake et al., 2023).

IoT technology devices make it easier for farmers to monitor the health of their soil by using sensors and analytical tools to measure nutrient concentrations. This data informs farmers how much to fertilize and where not to, helping prevent the overuse of chemicals that can be harmful to the environment. Soil monitoring integrated with PA can help farmers manage soil health and fertility, reducing overfertilization to improve crop yield.

8.3.4 Influence of Ambient Conditions

Plant growth rate and crop yield are influenced by all these parameters together with crucial ambient factors, such as temperature and humidity. Temperature impacts several plant physiological processes, including photosynthesis, respiration, and flowering. Every crop has an optimal temperature range within which it grows; anything above or below that threshold will decrease growth rates. Increased temperatures might speed up crop development but can also cause heat stress; decreased temperatures may reduce growth, resulting in frost damage (Kaswan et al., 2023).

Plant health is also influenced by humidity—also known as the quantity of dampness in the air. With high humidity, transpiration rates will be lower, potentially leading to the accumulation of excessive moisture on plant surfaces and increased instances of fungal diseases. On the other hand, with low humidity, water loss through transpiration is extreme; this induces stress in crops and limits growth. By using IoT sensors, farmers can easily monitor and maintain temperature and humidity levels to establish an ideal environment for plant growth, modifying greenhouse environments or irrigation practices—the choice is up to them.

8.3.5 Light 24-h Photoperiod and Light Intensity

Two important factors that influence plant growth and development include light intensity and photoperiod (length of daylight). Photosynthesis: Plants need light to make food through photosynthesis. Photosynthesis rates are affected by light intensity, and poor crop yields occur due to insufficient light.

Photoperiod refers to the amount of light and dark hours in a 24-hour period. Day length is responsible for inducing several developmental stages in plants, including flowering and fruiting. Certain crops need a specific photoperiod to flower. Farmers can understand more effectively how these parameters affect their crops. Furthermore, farmers can monitor light intensity and photoperiod by using IoT sensors to improve their farming methods or greenhouse conditions to increase plant growth and productivity.

8.3.6 Wind Speed and Direction

The influx of cool air is associated with variations in wind speed and direction, which can have direct and indirect effects on agricultural practices. Crops can also suffer physical damage from high winds—for example, lodging, where entire plants are blown over—reducing yields and crop quality. Wind can also affect the rate of evapotranspiration, or movement of water from soil and plants to the atmosphere, increasing irrigation water needs.

However, with the help of IoT sensors that monitor wind speed and direction, farmers can control and minimize these concomitant factors. For example, farmers will be able to use the data for siting and designing windbreaks or switching their irrigation schedules to account for potential increased evaporation. Wind data can be a valuable tool for farmers in adjusting their management practices to keep crops safe from wind-based damage while making the most of the water applied to irrigated fields.

8.3.7 Relationship between Precipitation and Evaporation

Both are crucial for the water the soil and crops have, through evaporation and precipitation, respectively. Precipitation is how water gets to plants in the first place and evaporation is the cycle of water out of the soil or off surface of plant up into the atmosphere. Stages of plant growth and development both interact with soil moisture, a critical variable for its overall production.

Too much rain can result in saturated soil, leading to root diseases and the effect on their general health. Conversely, low rainfall may lead to drought stress and decrease in crop yields while farming demanded irrigation. Soil moisture dynamics are a function of changes in temperature and evaporation rates (which are also linked to climate conditions), how much wind there is, and humidity.

Monitoring precipitation and evaporation by using IoT sensors are one of the most effective ways of managing water resources. By combining that with irrigation systems, farmers can then know when and how much to water their crops to provide them with the perfect amount of water. This is beneficial in balancing soil moisture levels, optimal use of water, and also enhances crop productivity (Table 8.1).

TABLE 8.1 Precipitation and evaporation

ASPECT	*PRECIPITATION*	*EVAPORATION*
Definition	The process by which water, in the form of rain, snow, sleet, or hail, falls from the atmosphere to the Earth's surface.	The process through which liquid water is converted into water vapour and rises into the atmosphere.
Phase change	Involves the transition of water from the atmosphere (gas) back to liquid or solid forms (rain, snow, etc.).	Involves the transition of water from a liquid state to a gaseous state (vapour).
Primary mechanism	Occurs when atmospheric moisture condenses and becomes too heavy to remain suspended, causing it to fall to the ground.	Occurs when heat energy is applied to water, increasing its temperature until it transforms into vapour.
Driving forces	Atmospheric cooling, saturation of air with moisture, and cloud formation trigger precipitation.	Solar radiation and wind speed are the primary drivers, increasing water's surface temperature to initiate evaporation.

(*Continued*)

TABLE 8.1 (*Continued*) Precipitation and evaporation

ASPECT	*PRECIPITATION*	*EVAPORATION*
Influencing factors	Cloud type and formation, humidity, atmospheric pressure, and geographic location (mountains, oceans, etc.).	Temperature, humidity, wind speed, water surface area, and altitude affect the rate of evaporation.
Occurrence	Can happen at any time when moisture-laden clouds are present and conditions are right. Common in humid, tropical regions, or mountainous areas.	More likely to occur in warm, dry climates or during periods of intense sunlight. Higher in areas with large bodies of water.
Types	• Rain, snow, sleet, hail. • Types depend on temperature and atmospheric conditions.	• Evaporation from oceans, lakes, rivers. • Transpiration (evaporation from plant leaves) is a related process.
Role in water cycle	Returns water from the atmosphere to earth's surface, replenishing water bodies like rivers, lakes, and oceans.	Moves water from earth's surface to the atmosphere, contributing to cloud formation and humidity.
Impact on environment	Precipitation replenishes freshwater sources, nourishes ecosystems, and influences agricultural productivity. Excess precipitation can cause floods.	Evaporation regulates earth's temperature, maintains humidity levels, and is key to cloud formation. Excessive evaporation can deplete water resources.
Measurement tools	Rain gauges, weather radar, and satellites.	Evaporation pans, atmometers, and satellite-based remote sensing systems.
Units of measurement	Measured in millimetres (mm) or inches of rain per given time period.	Measured in millimetres per day (mm/day) or kilograms per square metre (kg/m^2).
Effects of climate change	More intense and frequent precipitation events, potentially leading to flooding or prolonged droughts in some regions.	Increased evaporation due to rising global temperatures, leading to lower water levels in reservoirs, lakes, and other water bodies.

(*Continued*)

TABLE 8.1 (*Continued*) Precipitation and evaporation

ASPECT	*PRECIPITATION*	*EVAPORATION*
Examples	• Tropical rainstorms in equatorial regions. • Snowfall in polar or high-altitude regions.	• Evaporation from the surface of oceans and lakes. • Evaporation of water from plant leaves (transpiration).
Human Influence	Urbanization and deforestation can alter precipitation patterns by affecting local weather and increasing runoff.	Water management practices (e.g. irrigation) and climate change may accelerate evaporation rates and impact water availability.

8.4 IoT SENSORS AND DEVICES FOR SOIL AND AMBIENT MONITORING

IoT sensors are pivotal in monitoring soil and ambient conditions to optimize agricultural practices. Several types of sensors measure various aspects of environmental factors: Soil moisture sensors—measure the volumetric water content in the soil. They help in determining when to irrigate, minimizing water wastage, and ensuring the plants receive optimal levels of moisture. Common types include capacitive sensors that measure the dielectric constant of the soil and resistive sensors that detect changes in electrical resistance in terms of soil moisture. Temperature sensors—monitor the temperature of the soil and ambient air. Accurate measurements are crucial in determining the conditions for plant growth, predicting frost conditions, and optimizing heating and cooling in controlled environments. Thermocouples and thermistors are used due to their precision and reliability. pH sensors—measure the acidity or alkalinity of the soil, influencing nutrient availability and plant health. The soil pH determines the productivity of crops, and pH sensors utilize electrochemical methods to accurately measure pH. Other ambient sensors—in addition to soil sensors, multiple ambient sensors measure various other ambient conditions such as humidity, light intensity, wind speed, and barometric pressure. Measurement is essential in determining other broader climate conditions.

8.4.1 Sensor Deployment and Calibration

Accurate and reliable data result from effective deployment and calibration. Deployment—sensors are placed in a manner that data collected is representative of the area of interest. Soil moisture sensors, for example, are deployed in various depths to measure moisture levels throughout the root zone. Temperature and humidity senso are placed in areas that reflect as closely as possible natural environments, avoidin areas with heating elements or extreme shading.

Calibration: Regular calibration is necessary to ensure that sensors maintain th precision of their measurements. Calibration includes comparing sensor readings wit already known reference values, and correspondingly adjusting the sensors' settings. I the case of soil moisture sensors, the calibration may involve using sensor soil sampl with already known moisture values. Temperature and pH sensors are usually mair tained using measurements with the use of standard solutions or in conjunction wit another material with known values. Appropriate calibration mechanisms are critic to ensure that the data measured remains reliable and accurate throughout the life c the equipment.

8.4.2 Data Collection and Transmission

Data collection and transmission are the central elements of an IoT monitoring systen Data collection—sensors gather data always on different environmental condition sometimes continuously. This data is often collected regularly to provide a comprehen sive understanding of how conditions change over time. Advanced sensors could allo the collection of data in real-time, meaning that farmers receive alerts of changin conditions immediately. Data transmission—after collecting data, the sensor transmit the same to a centralized system for analysis. This is achieved using a Wi-Fi, cell phon network, or low-power wide-area network such as a LoRaWAN. The choice of networ to use will depend on how far the sensors are located, energy consumption, or hov frequent data transmission is required. Data management—the transmitted data is sen to the cloud or localized storage system for processing and analysis. IoT platforms offe the tools needed for visualization and parameter setting based on the data. This laye allows the farmer access to the data and to give instructions to the device. Data integra tion and analysis—integration of data from soil and ambient is crucial since data fron each set of sensors might not be enough to make the required decision. Several integra tion scenarios can be adopted:

Central Data Platforms: Central data platforms gather and combine sensor and data source information in a single database. Next, this aggregate gets integrated to provide grip on both soil and ambient conditions as a whole. The data from soil moisture sensors and temperature gauges and weather stations can provide a picture of the broader environmental situation influencing crop development.

Data Fusion Methods: In data fusion, information flows from different sensors as well as sources are integrated to provide an elaborate analysis. Several methodologies like multisensor fusion and data assimilation are being deployed to combine the information from soil sensors with weather information to provide a holistic insight into how the different variables interact and influence the growth of plants.

APIs and Data Connectors: Application programming interfaces (API) and data connectors allow to connect different systems and platforms. IoT sensors can automatically move data to the cloud platform that moves it from there to farm management software and vice versa by using APIs. It enables real-time data updates and universal app analysis by integration.

8.5 DATA ANALYSIS TOOL AND TECHNIQUES

Complex methodologies and tools are employed to derive useful insights from integrated soil and ambient sensor data, enhancing decision-making in areas such as agriculture, environmental monitoring, and civil engineering (Table 8.2):

Health Centre Implemented Three Machine Learning Techniques:

Descriptive Healthcare Analytics: This is to summarize historical data and provide visualization insights on past trends and patterns as well as clues in the understanding of how they affect certain elements, which make key members, agents, and roles. Data visualizations in the form of summaries, statistical results as well as charts help data to be presented in a way that helps farmers pick up on trends and anomalies of their soil and surroundings.

Diagnostic Analytics: Diagnostic analytics helps to understand why a certain outcome occurred (or happening) by checking the leading dimension of those outcomes. They can figure out what was causing the issues such as poor crop growth or water stress by analysing data correlations and relationships. It will help analysts to respectively establish the correlation as well as identify the root cause of specific issues.

Prediction Analytics: Prediction analytics also takes on historical and real-time data to anticipate future conditions and outputs. Examples of target-related variables include crop yields, pest outbreaks, and when irrigation would be most optimal to reduce the amount of water used. Machine learning algorithms—by being better prepared with predictive models, farmers are able to not only manage the existing conditions around them but start to plan and resource as needed for what is on the horizon.

Prescriptive Analytics: The result of prescriptive analytics is recommendations, often very specific about what should be done next to solve a problem or improve upon a situation. These models give their output through a few optimization algorithms or decision-making support systems regarding specific actions concerning certain applications such as how much to irrigate, when to fertilize, and also which pest control measures must be employed. This sort of analysis can assist in noticing improvement opportunities for farmers in making smarter decisions based on data to increase the performance of crops and resource efficiency.

TABLE 8.2 Data analysis tools and techniques

TOOL/TECHNIQUE	*DESCRIPTION*	*USE CASE/ APPLICATION*	*STRENGTHS*	*LIMITATIONS*
Python (with Pandas, NumPy)	Python is a versatile programming language with libraries like Pandas and NumPy, commonly used for data manipulation, cleaning, and statistical analysis.	Data cleaning, manipulation, exploratory data analysis, and statistical modelling.	Open-source, large community support, flexibility, and a rich ecosystem of libraries for data science.	Requires coding knowledge; can be slower compared to specialized tools for very large datasets.
R	R is a programming language and environment specifically designed for statistical computing and graphics.	Statistical modelling, hypothesis testing, machine learning, and visualizations.	Specialized for statistical analysis, excellent for exploratory data analysis, and rich library of packages (e.g. ggplot2).	Steeper learning curve for beginners; less versatile than Python for general-purpose programming.
Tableau	A powerful data visualization tool that enables users to create interactive and shareable dashboards from various data sources.	Business intelligence, real-time data visualization, and reporting.	Easy-to-use, drag-and-drop interface, powerful visualizations, integrates with many data sources.	Limited advanced statistical or machine learning capabilities; can be costly for large organizations.
Excel (with Power Query)	Microsoft Excel, with Power Query, provides tools for data analysis, transformation, and visualization, useful for non-technical users.	Quick data analysis, data cleaning, and small to medium datasets.	Widely accessible, easy-to-use interface, strong support for smaller datasets and common tasks like pivot tables.	Limited handling of large datasets; lacks advanced statistical and machine learning features; can be slow for complex tasks.
Apache Hadoop (with Hive, Pig)	A framework for distributed storage and processing of large datasets across clusters of computers using simple programming models.	Big data analytics, data storage, and processing in distributed systems.	Handles large-scale data, fault-tolerant, and integrates with other big data tools like Hive, Pig, and Spark.	Complex to set up and manage; requires technical expertise in big data systems and infrastructure.

(*Continued*)

TABLE 8.2 (*Continued*) Data analysis tools and techniques

TOOL/TECHNIQUE	*DESCRIPTION*	*USE CASE/ APPLICATION*	*STRENGTHS*	*LIMITATIONS*
SQL (Structured Query Language)	SQL is used to manage and manipulate relational databases. It allows for querying, updating, and analysing structured data stored in databases.	Data extraction, querying structured data, and relational database management.	Highly efficient for querying large relational datasets; widely used and supported; integrates well with other tools.	Limited for unstructured data; requires knowledge of database structures and SQL syntax.
Apache Spark	An open-source unified analytics engine for large-scale data processing, supporting in-memory computation.	Real-time data processing, machine learning, and big data analytics.	Fast processing of large datasets in distributed environments; supports machine learning and stream processing.	Requires infrastructure setup and expertise; not as flexible for small datasets or simple analysis tasks.
MATLAB	A programming platform designed for engineers and scientists to analyse and design systems and products.	Signal processing, statistical analysis, and algorithm development.	Excellent for numerical analysis, built-in functions for complex mathematics, strong support for engineering tasks.	Commercially licensed; less commonly used in non-engineering fields; not as flexible as Python for general use.
SPSS (Statistical Package for the Social Sciences)	A software suite used for statistical analysis in social science, healthcare, and market research.	Descriptive statistics, hypothesis testing, regression, and data mining.	Easy-to-use interface for statistical analysis, well-suited for social science data, strong for survey data analysis.	Can be expensive; less flexible than Python or R for advanced custom analyses; and limited machine learning functionality.
Power BI	A business analytics service by Microsoft that delivers insights from data by enabling data visualization and reporting.	Business intelligence, real-time data visualization, and reporting similar to Tableau.	User-friendly interface, integrates well with Microsoft products, excellent for business analytics and reporting.	Limited statistical and machine learning capabilities; can be expensive for large-scale or enterprise applications.

8.6 REAL-TIME MONITORING AND PREDICTIVE ANALYTICS

Modern agricultural practices depend a lot on real-time monitoring and predictive analytics as it allows farmers to react faster to changes as well as predict problems beforehand as shown in Figure 8.1.

Real-Time Monitoring: Real-time monitoring involves continuous monitoring of soil conditions and other ambience parameters by utilizing IoT sensors. This allows quick feedback about such environmental changes such as soil moisture variations or abrupt temperature changes. With real-time data, farmers can access information instantaneously and make necessary changes—like turning on irrigation systems or adjusting ventilation in greenhouses—to keep growing conditions ideal.

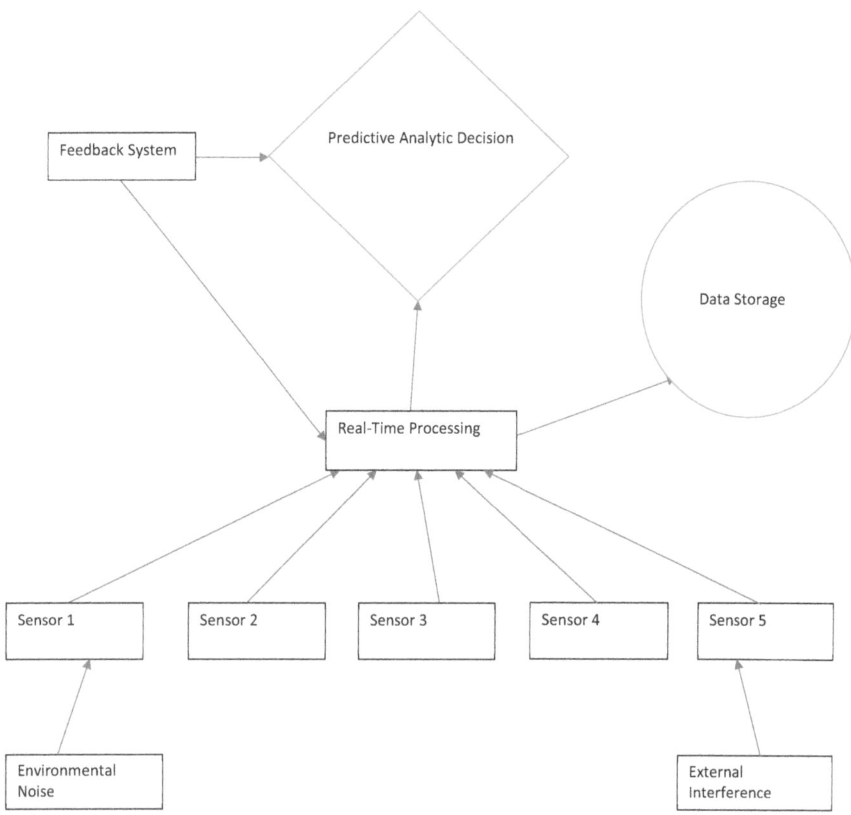

FIGURE 8.1 Real-time monitoring.

Predictive Analytics: A predictive analysis model uses real-time data and historical information to predict future states or events. To name a few, data-driven predictive models can be utilized by farmers today to determine future soil moisture conditions—hence optimize when to irrigate and how much irrigation is needed based on current weather patterns along with historical information. Moreover, predictive analytics is also capable of predicting potential pest infestations, or disease outbreaks so preventive measures can be taken to save the crop from getting destroyed.

Alerts and Notifications: A few useful monitoring systems enable certain alerts to be set upon farmer need too, if essential conditions are violated. These can be predetermined, for example, to determine the possibility of low soil moisture, high temperature, etc. These alerts allow farmers to react quickly and address potential problems before they arise, thus maintaining crop health.

8.7 CASE STUDIES IN PRECISION AGRICULTURE

IoT-enabled soil and ambient monitoring represents just one of the many known implementations in which applying IoT technology to agriculture continues to show its worth.

CASE STUDY 1: PRECISION IRRIGATION IN CALIFORNIA

A large farm near California's Central Valley is providing an example of a commercial-scale application of an IoT precision irrigation system that adjusts water usage based on data from the field. Real-time soil moisture penetration probes were installed at different depths to provide information for soil water content. This data was coupled with long-term weather forecasts and crop growth models to adapt irrigation scheduling in real-time. A 30% decrease in water use also meant a 15% increase in crop yields as irrigation became more efficient.

CASE STUDY 2: SMART GREENHOUSE MANAGEMENT, THE NETHERLANDS

We could use the example of an IoT scenario where we monitor environmental conditions in a greenhouse operation that would be somewhere in the Netherlands. Systems automatically monitored temperature, humidity, and light via sensors, with ventilation, heating, and shading then controlled based on real-time data. Using this technique allowed them to increase plant growth rates by 20% and reduce the costs of controlling the climate in energy by 25%.

CASE STUDY 3: GRAPE GO FRENCH VINEYARD MANAGEMENT

French vineyards used IoT sensors to measure soil moisture, temperature, and vineyard microclimates. This data was used to introduce precision viticulture, by controlling water and nutrient supply according to an exact prescription based or sensor feedback. This made it possible to achieve the technological organization of wine production, improve the quality of grapes, and increase the efficiency of wine production by 10%.

8.8 IMPLEMENTATION OF IoT TECHNOLOGY

When IoT technology is applied to agriculture, several valuable lessons are learned an insights are provided that can pave the way for future projects.

Integration and compatibility—when working with many sensors and systems it is important to make sure that all of the elements are integrated together well. I other words, plugging in with various IoT devices and software platforms achieved th interoperability necessary for tracking as a whole. Choose devices and technologies tha align effectively and integrate easily into your ecosystem.

Calibration and Maintenance: Pressure sensors must be calibrated at regula intervals to ensure that the readings are accurate. Non-calibrated or improperly func tioning sensors can cause wrong data, which will in turn affect decision-making i agricultural outcomes. Default sensor calibration and maintenance routine contribute t proper data accuracy and system reliability.

Similarly, to work with IoT sensors producing that vast amount of data, you nee to have a better data management style. Strong data storage solutions ensure secure an efficient management, with data broken down into bite-sized pieces for faster process ing, easy retrieval, and improved scalability; well-secured data and fast parallel/distrib uted/anionic processing analysis tools are a must-do to derive the best outcome out o your existing collective information. Clear data management protocols ensure that dat continues to be accurate and accessible.

8.8.1 Comparative Study with Different Techniques

Each approach of IoT implementation in agriculture has its own strengths and factor that need to be taken into account, according to the comparative analysis.

8.8.1.1 Method 1: On-premises vs cloud-based systems

The on-premises option is for installing and maintaining IoT infrastructure locally giving more control and less latency. However, they might be expensive to build and maintain on the front end. Meanwhile, cloud-based solutions offer scalable options with the benefits of wide accessibility and lower costs for infrastructure. They are perfect for farms with very limited on-site resources but a good internet connection.

8.8.1.2 Approach 2: Passive vs active monitoring systems

This includes data that is not collected by actuating (for example, static soil moisture sensors) or in other words through passive monitoring systems. While these systems provide useful insights, any corrections will require manual oversight. Passive monitoring systems provide average values over time that serve for general system tuning, whereas more dynamic control methods such as automated irrigation systems adapting in real-time according to sensor data may be of higher implementation cost and complexity.

8.8.1.3 Technique 3: Single vs. multisensor systems

A single-sensor system may target a desired parameter like soil moisture or temperature. While these methods are less complicated and of lower cost, they do not offer a holistic view of real environmental conditions. Multiple sensors, which are the combined signals from different areas, lead to a solution with a more comprehensive dimension because of gathering diversified data in nature. While this provides a level of better decision-making and more up-to-date information in the process, it is significantly costlier and more complex.

8.8.2 Challenges and Restrictions

The most tenacious challenge in adopting an IoT-based sensor solution for soil and ambient monitoring is to provide highly accurate and precise data from the sensors.

Some Common Bottlenecks: Calibration issues—to ensure accurate readings when using sensors, they must be calibrated accurately. Calibration mismatches can also introduce inaccuracies in readings, which results in the development of wrong conclusions. For example, soil moisture sensors can drift over time and need to be frequently recalibrated to preserve accuracy.

Sensor Drift/Degradation: Environmentally related factors including extreme temperatures or exposure to chemicals can cause drift and degradation of the sensors over time. This can make long-term data collection less reliable, and require it to be more frequently serviced or replaced.

Environmental Inference: Environmental conditions like high humidity, dust content, or chemical residues may make it difficult for sensors to perform their functions properly. One such instance may be when soil pH sensors register wrong results because of the products made use of for fertilization, or it may occur due to harsh soil conditions.

Precision Limitations: Sensors do not have the same precision. Some sensors may have highly accurate readings, whereas others are more general. Selecting and specifying the correct sensor has to fit the specific requirements of the application in order that these highest standards are met.

8.8.3 Data Management and Processing Issues

Challenges: However, the management and processing of data that will be generated by these IoT sensors are cumbersome.

Volume and Storage: The large volume of data generated by IoT systems can become unmanageable both in the amount to process and store. To accommodate this wealth of information without corrupting the data, efficient data storage solutions and data compression techniques are necessary.

Data Integration: Integrating data across multiple sensors and sources can be challenging. Differences in data formats, communication protocols, and data structures are obstacles to integration without a lot of work. We need to find data fusion techniques that blend different datasets together for a comprehensive analysis.

Data processing, analysis, and analysing big datasets demand heavy-duty computational resources along with sophisticated analytical tools. Processing the data quickly enough and turning that data into insights can also be difficult when real-time analytics are required by organizations due to a lack of computing power or using previously outdated software.

Security and Privacy: It's of utmost importance to protect your data from unauthorized access and ensure privacy. Cyber threats and data breaches are common risks for IoT systems. Protecting sensitive information requires that you follow best practices in terms of security, for example, data encryption and access control.

The various environmental and operational factors that directly affect IoT deployment in agriculture are as follows:

Longevity and Environmental Conditions: Harsh environments such as very high or low temperatures, high humidity, or rain can reduce the longevity of sensors as well as their performance. For one thing, sensors buried in the soil can be thrown off by changing moisture levels or temperature changes, which skew their accuracy.

Power Supply: Maximum IoT sensors and devices are powered by electricity. Delivering power without reliable grid access can be equally challenging for remote or off-grid locations. This limitation requires solutions such as solar-powered sensors or energy-efficient devices.

Connection: Sensors must deliver the data to central systems; hence, a reliable connection is necessary in this situation. These files may only be uploaded in locations with adequate bandwidth and reliable connectivity, though this is often not an option for many rural or remote areas where cell signal is terrible and internet availability is low. It is critical to carry out a steady communication medium for proper real-time monitoring and data transmission, which helps organizations with better visualization.

Operational Complexity: Setting up and maintaining an IoT monitoring-based solution can be complex and it requires specialized knowledge. Users need training and support to correctly use the system on a daily basis.

8.9 FUTURE TRENDS AND INNOVATIONS

Over the coming years, a number of new technologies are expected to have transformative effects on agriculture, particularly in the following areas:

Advanced Sensor Technologies: New versions of sensors are under development, which can provide enhanced precision and strength. An example of this is the nano and smart sensor that can accurately detect small changes in soil and environmental conditions. These can also incorporate more than one measurement capability in the same device, cinching the requirement for multiple sensors.

What Comes Next: Remote sensing and drones—more commonly used remote sensing tools include drones with multispectral and hyperspectral cameras. These drones can cover large tracts of agricultural land from the air, delivering information on crop health, soil, and environmental conditions in detail. Thanks to the drones, high-resolution images and data they collect help for real-time analysis and decision-making.

Wireless Sensor Networks: WSNs allow the installation and integration of hundreds to thousands of networked sensors over a geographical area, allowing extended coverage with minimal cabling. This assists with the real-time monitoring and data collection over multiple locations, aiding in the management and analysis of soil as well as ambient conditions in real-time.

Functionalities of Biological Sensors: A biological sensor is a device that can be employed to sense an analyte by using living organisms like microbes or plants themselves. By using the processing of natural biological organisms, these sensors can also give insight into soil health and nutrient levels.

8.9.1 IoT Leads the Agriculture Industry

Advancements of IoT technology has made it possible to have new opportunities in agriculture.

Edge Computing: Edge computing provides the processing of data to some extent close to its source generation, enabling reduced latency and faster response times. For the agriculture industry, edge computing can help deliver real-time sensor data analysis at site level to be able to act instantly based on actual conditions without being slave anymore by cloud systems.

Artificial Intelligence (AI) and Machine Learning (ML): AI and ML are getting built into IoT systems more in order to amplify data analysis techniques along with the predictive capabilities. These tools can capture insights, analysing big data sets and detecting the hidden patterns. AI algorithms can, for instance, predict soil-water requirements or pests in the case of historical and time-series derived from data.

Blockchain for Data Security: Blockchain is used to secure agricultural data and provide integrity. With a decentralized and immutable ledger, blockchain serves as an effective tool for eliminating data tampering and enhancing transparency in the process of managing the information.

Automated Systems and Robotics: Automation and robotics have been increasingly used in agriculture, with the creation of systems capable of helping to perform planting, weeding, and harvesting based on IoT data. They increase efficiency, eliminate or reduce manual labour, and are aimed at improving productivity in agriculture.

Improved Integration and Interoperability: The advancements to come will be geared towards building more integrated systems that are interoperable with respective devices, sensors, and platforms. This will allow a unified manner of collecting and analysing data that will enable better-informed decisions.

Increasing Data Analytics and Visualization: The combination of improved sensor data, better data analytics, and visualization tools will give farmers more powerful new ways to understand what the data is telling them. More sophisticated forms of visualization, such as 3D mapping and augmented reality will start to allow farmers to see the ground beneath them better than ever before helping from understanding field histories, to making in-situ decisions on what and where to plant.

Increased Focus on Sustainability: With sustainability becoming an integral theme in agriculture, IoT technology will ensure that more sustainable practices are pursued. This pillar of innovation will serve to help farmers optimize their resource utilization, reduce waste, and grow healthy soil, all while balancing sustainability with efficiency, ultimately advancing resilient and eco-friendly farming practices.

Bespoke Farming Solutions: As we move into the future, connected farming solutions will become more bespoke to individual farms and crops—providing unique IoT systems that match exact needs. Customizable solutions that specifically cater to farmers' unique challenges and needs will allow them to use best practices for their exact conditions.

Global Supply and Demand: The expansion of the IoT among agriculture services will reach untapped areas across the globe. These endeavours will target at reducing cost and ease of use barriers, making IoT solutions available and affordable to a wide variety of farmers in poor countries with the overall bio-productivity improvements participating to more general common uplift.

8.10 CONCLUSION

Indeed, the advancement of technologies and their incorporation into various segments of agriculture led to a rather precise decade. This chapter focussed on the importance of soil and environmental variables in the progress of precision farming, enlightening how deeply they affect crop productivity. We have covered the basics of IoT that are, what are its building blocks, advantages, and some of its challenges coming to an agricultural perspective how can this stack (IoT) change traditional farming practices. Case studies also demonstrate what successful implementations of IoT look like for soil monitoring and ambience monitoring, respectively, providing valuable insights into best practices while observing the relative advantages of various tricks. However, along with these opportunities, a number of challenges remain around the considerable sensor accuracy requirements, robust data management, and very difficult environmental working conditions. The future of PA is bright, with new technologies and innovations on the horizon that promise to improve IoT functionality. While we progress, comprehensive R&D is required to overcome the current infirmities and pave the way for better future prospects with the choice of agricultural acumen. In addition, with that, we come to a wrap-up point in this chapter, one that suggests how the IoT can intervene potentially altering the way agriculture occurs and bring about sustainable and productive farming.

REFERENCES

Akhter, R., & Sofi, S. A. (2022). Precision agriculture using IoT data analytics and machine learning. *Journal of King Saud University-Computer and Information Sciences*, *34*(8), 5602–5618.

AlZubi, A. A., & Galyna, K. (2023). Artificial intelligence and internet of things for sustainable farming and smart agriculture. *IEEE Access*, *11*, 78686–78692.

Castrignanò, A., Buttafuoco, G., Khosla, R., Mouazen, A., Moshou, D., & Naud, O. (Eds.). (2020). *Agricultural Internet of Things and Decision Support for Precision Smart Farming*. Academic Press.

Chithaluru, P., Singh, A., Dhatterwal, J. S., Sodhro, A. H., Albahar, M. A., Jurcut, A., & Alkhayyat, A. (2023). An optimized privacy information exchange schema for explainable AI empowered WiMAX-based IoT networks. *Future Generation Computer Systems*, *148*, 225–239.

?J

Dhanaraju, M., Chenniappan, P., Ramalingam, K., Pazhanivelan, S., & Kaliaperumal, R. (2022 Smart farming: Internet of Things (IoT)-based sustainable agriculture. *Agriculture*, *12*(1C 1745.

Fuentes-Peñailillo, F., Gutter, K., Vega, R., & Silva, G. C. (2024). Transformative technologies digital agriculture: Leveraging Internet of Things, remote sensing, and artificial intelligen for smart crop management. *Journal of Sensor and Actuator Networks*, *13*(4), 39.

Goel, R. K., Yadav, C. S., Vishnoi, S., & Rastogi, R. (2021). Smart agriculture–Urgent need of tl day in developing countries. *Sustainable Computing: Informatics and Systems*, *30*, 10051

Kalyani, Y., & Collier, R. (2021). A systematic survey on the role of cloud, fog, and edge compu ing combination in smart agriculture. *Sensors*, *21*(17), 5922.

Karunathilake, E. M. B. M., Le, A. T., Heo, S., Chung, Y. S., & Mansoor, S. (2023). The path smart farming: Innovations and opportunities in precision agriculture. *Agriculture*, *13*(8 1593.

Kashyap, B., & Kumar, R. (2021). Sensing methodologies in agriculture for soil moisture ar nutrient monitoring. *IEEE Access*, *9*, 14095–14121.

Kaswan, K. S., Dhatterwal, J. S., Baliyan, A., & Rani, S. (2023). *Quantum Computing: A New E of Computing*. John Wiley & Sons.

Kaswan, K. S., Dhatterwal, J. S., Ojha, R. P., Balusamy, B., & Gangadevi, E. (2024). AI-base AR/VR models in biomedical sustainable Industry 4.0. In E. Gangadevi, M. Lawany Shri, Rajesh Kumar Dhanaraj, Balamurugan Balusamy (Eds.), *Computational Intelligenc in Bioprinting* (pp. 53–78).

Placidi, P., Morbidelli, R., Fortunati, D., Papini, N., Gobbi, F., & Scorzoni, A. (2021). Monitorin soil and ambient parameters in the IoT precision agriculture scenario: An original mode ling approach dedicated to low-cost soil water content sensors. *Sensors*, *21*(15), 5110.

Shaikh, F. K., Karim, S., Zeadally, S., & Nebhen, J. (2022). Recent trends in inte net-of-things-enabled sensor technologies for smart agriculture. *IEEE Internet of Thing Journal*, *9*(23), 23583–23598.

9 Revealing and Location of Rock Falls via Seismic and Infrasound Measuring Device

9.1 INTRODUCTION

Rock falls are a common geotechnical hazard, particularly in mountainous terrains and steep slopes. The sudden movements of rock can create destruction to infrastructure, pose a very high level of hazard for people, and disrupt transportation networks along with other critical systems. Thus, the rock fall study is very important for geotechnical engineering to comprehend these processes and work out mitigation measures. The sound produced by rock falls strictly relies on fast detection and precise localization of the events to allow timely intervention before important damages take place (Zimmer & Sitar, 2015).

9.1.1 Rock Falls in Geotechnical Engineering

Within geotechnical engineering, rock falls are among the most unpredictable natural occurrences. That happens when loose rocks loosen from slopes due to natural erosion, freeze-thaw cycles, or human activities such as mining and construction. For civil engineers rooted in areas where rock fall hazards are more common, understanding why rocks fall can be vital. On top of the danger they pose in their own right, rock falls can

DOI: 10.1201/9781003537625-9

trigger other forms of peril (e.g. landslides or avalanches) that further increase hazards to life and property. Therefore, the main geotechnical work in relation to rock fall prevention and detection is crucial (Collins & Znidarcic, 2004).

9.1.2 The Significance of Detection and Localization (Collectors)

Real-time rock fall detection and localization (Figure 9.1) represent a paradigm shift for geotechnical hazard management. We also have to consider the value of early detection: protecting infrastructure and saving lives thanks to being able to evacuate or take necessary actions before it is too late. Rock falls can occur unpredictably, so real-time monitoring is necessary (Deparis et al., 2008). To protect large sections of the population from these potentially catastrophic failures, geotechnical engineers need their best technology to identify even the tiniest hints that such an event is imminent—a hint as ephemeral as distorted ground vibrations or acoustic resonances. This allows for targeted responses in particular geographic locations, reducing the overall burden of these events (Ulivieri et al., 2020).

9.1.3 Problems with Detecting Rock Falls

Several technical obstacles need to be overcome in the detection of rock falls, mainly originating from the different nature of the ground and sporadic settlement. However, traditional visual monitoring systems, such as manual inspections or surveillance cameras, are often insufficient, especially in remote and hard-to-reach areas. Small or slow-moving falls might seem unimportant but can be dangerous in vast numbers. This in turn requires the creation of sophisticated automated detection systems capable not only of round-the-clock surveillance across vast territory but doing so under harsh environmental conditions (Schmandt et al., 2013).

9.1.4 Rock Fall Detection Using Seismic Technologies

Recently, seismic methods have become the most successful way to detect rock falls. The technology uses sensors placed on the ground to detect vibrations from rocks falling at various calibration levels. Seismic monitoring stations record even relatively small rock falls and can provide information on the exact timing, size, and source location of an event. Engineers can then analyse the seismic signal and determine if rock falls raise similar seismic alerts to those triggered by earthquakes, helping improve their monitoring abilities (Xie et al., 2023).

9.1.5 Rock Fall Detection Using Infrasound Technologies

Infrasound monitoring provides a second tool for detecting rock falls. Mammals and birds can detect infrasound, or sound waves with frequencies below the lower limit of human audibility. As rocks tumble down slopes, they emit infrasound sound waves with frequencies lower than the range of human hearing, and to detect these low-frequency signals, scientists installed specialized sensors. While seismic sensors record ground vibrations, infrasound sensors pick up those in the air, making them suitable for tracking over long distances. Seismic and infrasound data together offer a more complete record of rock fall events (Zimmer & Sitar, 2015).

9.1.6 Why Use Seismic and Infrasound Technologies

The main goal in rock fall monitoring through the means of seismic as well as infrasound technologies is to assure real-time detection and locating of emerging events. In doing so, though, the technologies are complementary: While seismic sensors "read" ground-surface movement, infrasound detectors can detect air pressure waves. Combining the two creates an integrated system that can detect all sizes of wet and dry rock falls during the night, day, or severe weather conditions. It uses a dual approach to ensure both the tiniest of rock falls can be spotted, barring those in motion at these dearth-of-motion frequencies almost immediately (Ulivieri et al., 2020).

9.1.7 EWS Enhancement

The early warning for rock falls can be significantly enhanced by one of the main objectives that enhances the integration of seismic and infrasound technologies. Thanks to their ability to detect events in real-time and provide accurate location data, these systems are capable of triggering alarms that alert authorities and stakeholders who can take action (Amitrano et al., 2005). Early warning systems, like those used in areas where rock falls are a constant danger, could make the difference between a minor event and a large disaster. This also plays a key role in supporting superior risk assessment models, which help engineers to better target high-risk rock fall areas with resource allocation (Schmandt et al., 2013).

9.1.8 How Data Helps to Remove Rock Fall Detection

Seismic and infrasound sensors have proven to be invaluable tools in detecting both airborne and rock-based phenomena. By analysing the large datasets collected by these systems, one can also identify commonalities such as annual changes in rock fall movements and patterns likely related to external environmental factors like rainfall or temperature changes. More advanced data analytics, including machine learning, can further enhance the accuracy of detection by discovering subtle signs for an impending rock fall event. In addition to increasing the current monitoring capability, an empirical data-driven approach could lead to better insights for further research and the development of more effective mitigation actions (Xie et al., 2023).

9.2 FUNDAMENTALS OF ROCK FALLS

Rock falls are sudden, spontaneous natural movements of rock masses, which detach from cliffs and steep slopes and travel rapidly downslope. Rocks falling directly downwards or bouncing off cliff and hill side which reservations landslides (downhill movement of soil/debris) that crowds are prone to mix it with them, as both rocks fall from the top of a slope. Falling rocks are either small individual stones or larger boulders and can come either in the form of separate entities or large rockslides. Minor rock falls are smaller in size, while major rock falls may have a volume of thousands of cubic metres taking place at a velocity high enough to make detection difficult as its volume approaches ultra-fast abrupt collapse (Table 9.1).

9.2.1 Causes and Triggers

Rock falls are initiated by a broad range of causes that may or may not be natural but there are both natural processes and human-induced factors. The weakest link in the house of diamond is weathering, a natural effect on a rock causing its breakdown into fragments similar to how erosion, earthquakes, and volcanic activity weaken the ground (Dietze et al., 2017). Repetitive freeze-thaw cycles are another common cause of rock breakup, as water in the cracks in rocks freezes and expands, eventually prising the rock apart. Moreover, heavy rainfall accelerates the porosity of slopes, thus weakening the friction bond maintaining rocks at equilibrium. Similarly, rock falls can occur due to the destabilization of slopes either through natural geomorphological phenomena (earthquakes) or by anthropogenic means such as mining, road construction, and deforestation. Vibration from heavy machinery or blasting can encourage the movement of an already marginally stable rock mass (Zimmer & Sitar, 2015).

9.2.2 Effects on Infrastructure and Safety

Rock falls can be a serious hazard to both infrastructure and human safety, particularly in regions that have steep slopes or mountainous terrain. For instance, highways, railways, tunnels, and bridges constructed in those areas are continuously threatened to be destructed. Falling rocks often block roads, leading to traffic snarls and collisions, and the rocks can also damage essential infrastructure like power lines and pipelines

TABLE 9.1 Basics of rock falls

FIELD	*DESCRIPTION*	*KEY CONCEPTS*	*MATHEMATICAL/ TECHNICAL ASPECTS*	*EXAMPLES/CASE STUDIES*
Causes of rock falls	Rock falls typically occur due to natural weathering processes such as freeze-thaw cycles, heavy rainfall, erosion, or seismic activities. Human activities like mining, construction, and road cutting can also trigger rock falls	• Weathering seismic activity • Erosion • Human interventions	Seismic force calculations	• Rock falls in mountainous regions due to freeze-thaw cycles in Switzerland • Mining-related rock falls in South Africa
Mechanisms of rock falls	The process involves the detachment of rock fragments from a steep slope, followed by freefall, bouncing, or rolling down the slope. The mechanism includes both gravity-induced forces and the material's structural weaknesses	• Gravitational force • Slope stability • Fracture propagation	Kinematic analysis	• Case of rock fall disasters on the Yosemite Valley cliffs in the US • Kinematic slope stability studies in the Alps
Effects of rock falls	Rock falls can have destructive effects on infrastructure, roads, and ecosystems, leading to property damage, injury, or loss of life. Rock falls also modify the landscape by depositing debris and altering drainage patterns in mountainous areas	• Impact force • Damage to infrastructure • Ecosystem changes	Impact energy calculation	• Impact of rock falls on roads in the Himalayas causing blockage • Damage to power lines in rural Japan due to rock falls

(*Continued*)

TABLE 9.1 (*Continued*) Basics of rock falls

FIELD	*DESCRIPTION*	*KEY CONCEPTS*	*MATHEMATICAL/ TECHNICAL ASPECTS*	*EXAMPLES/CASE STUDIES*
Monitoring of rock falls	Monitoring methods include field observations, remote sensing, laser scanning (LiDAR), and ground-based radar. These techniques help detect precursors to rock falls, assess slope stability, and predict potential rock falls	• Remote sensing • LiDAR scanning • Ground-based radar	Early warning models Statistical or machine learning models such as logistic regression to predict the likelihood of rock fall occurrence based on environmental conditions	• LiDAR-based monitoring of the Mont Blanc massif in France • Remote sensing studies in the Canadian Rockies to predict rock fall hazards
Mitigation strategies	Mitigation measures include rock fall barriers, nets, slope reinforcement, vegetation planting, drainage improvement, and controlled blasting. Proper geotechnical assessment and engineering interventions can help reduce risks in high-risk areas	• Rock fall barriers • Slope reinforcement • Vegetation • Controlled blasting	Barrier design: Dynamic rock fall simulation models to calculate required barrier strength based on rock size, fall velocity, and impact force	• Rock fall barrier installations in Norway along major roads • Slope reinforcement and netting in Italy's Dolomites to protect hiking paths and villages

(Walter et al., 2012). The continued impact of rock falls extends beyond the continuum of costs, loss, and disruption of transport; it is also a direct threat to human lives, particularly in areas where people simply live or work too close to unstable slopes (McNutt et al., 2015). The physical cost is immediate, and repairs, maintenance, and mitigation measures are very expensive; rock falls can be an important economic issue not to underestimate.

9.2.3 Weathering and Rock Fall

There is no denying that weathering is one of the most important processes in determining the development of rock falls. Rocks experience chemical, physical, and biological weathering over the years that disintegrates them. Mechanisms for breaking down rocks and minerals through bodily cycles such as thermal expansion, thawing of ice, or model volume recession may encourage physical weathering for behearkens which embody the typical rock for a given region. Chemical wear generally involves the mineral composition of rocks that reduces their grain size, resulting in a more microcoria texture into fragments; a process of liberation and capital is m thermal different in extract lie safe predominantly an emergent marriage degradation. Rocks expand and contract under thermal weathering cycles, and this also causes cracks in the rocks due to freeze-thaw action. Plants growing into rock crevices, among other biological factors, account for the further loosening of rock masses. Rocks start to become unstable and prone to falling as weathering continues (Feng & Zhuang, 2021).

9.2.4 Earthquake Can Be a Reason for Rock Falls

One of the most important natural factors that can trigger rock fall is earthquakes. Ground shaking from this seismic activity can also result in the falling of rocks from steep slopes or cliffs. In fractures and joints that already exist, even small tremors can cause large rock falls. This disaster is not limited to the primary rock fall due to a seismic event; rather, it can be induced in the coming days or weeks following several post-seismic rock falls. The natural drop moment often coincides with strong earthquakes in well-known seismic active areas such as the Pacific Ring of Fire, which is a further hazard for vulnerable regions and infrastructure.

9.2.5 Gradual Erosion over Time and Rock Falls

Rock falls are the result of slow but sure, long-term erosion processes through wind, water, and time—mostly now formerly glaciated mountains. Cliffs and slopes are undercut by erosive processes at their base over time as they erode supporting material, which can lead to the collapse of rocks resting upon it. In addition, as the softer rock erodes at the bottom, everything unsupported either swings or falls. Water erosion at the base of rock formations and along coastlines and riverbanks, where water is constantly washing away the rock below. Although these processes act over very long-time scales, they are highly significant in landscape evolution and the initiation of rock falls (Yan et al., 2020).

9.2.6 Rock Falls from Cliff: The Safety Implications in Residential Areas

In residential areas, such as places constructed based on or parallel to hills and mountain slopes, rock falls can directly threaten human life. Unstable slopes include cliffs, hills, and slide areas that can damage or destroy homes, schools, and other buildings. Large boulders can crash through buildings, causing injury or death. The danger in rock falls comes from their unpredictability—these events cannot be forecast, and residents may have only moments to take cover before a dynamic cascade pours through. For this reason, to prevent damage from rock fall events during the construction and development of residential areas, it is very important that geotechnical engineers and urban planners make assessments of rock fall hazards.

9.2.7 Rock Fall Risk Mitigation Measures

Engineers use several mitigation measures to reduce the risks of rock falls. This can be done by installing barriers, mesh, and embankments to stop the rocks from sliding down onto susceptible areas. Stabilization might become a reality as it will involve unique maintenance and rock bolting to keep any unstable rocks secure preventing them from falling. Warning systems that monitor ground vibrations, detecting early signs of rock movement, can be installed in high-risk areas, giving time for evacuation or other safety devices. Preventing rock falls is the optimal approach, and a mix of engineering solutions, detailed planning—like a pavement inspection schedule and a properly prepared maintenance plan—along with regular monitoring, are essential steps to protect people and roads.

9.3 GLOBAL SEISMIC MONITORING PRACTICES

This process consists of capturing ground vibrations created by the origin itself, varying from natural seismic events and human-made induced sources to mass movements (rock falls). Seismic monitoring is possibly one of the most easily understood principles, measuring energy waves that are propagated through the Earth's crust when an event occurs. These waves, primarily P-waves (primary waves) and S-waves (secondary waves), are detected by seismic sensors and measured, or recorded for analysis.

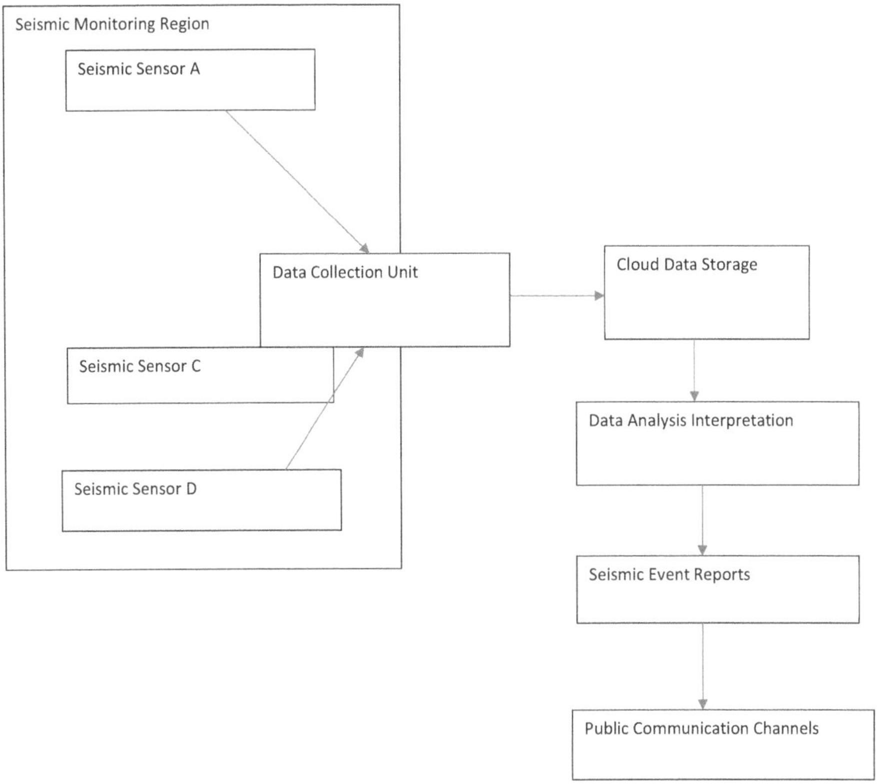

FIGURE 9.1 Global seismic monitoring.

The data collected reveals important details on the several different aspects of a seismic event including the time, place, size, and nature of it. Seismic monitoring can be used to identify the shock waves that accompany a rock fall, and hence quickly pinpoint where the triggering event occurred if it was based on a rockslide or topple failure. The detection of seismic information may also provide an early warning for slope stability issues as they are happening in Figure 9.1.

9.3.1 Types of Seismic Sensors

There are several types of seismic sensors used to monitor and detect rock falls, suitable for different terrains and applications. Geophones have been the cornerstone of seismometers for years, and remain one of the most popular sensors used in seismic monitoring. The devices, which are used to record the vital data on the intensity of earthquakes, convert the mechanical energy of ground motion into electrical form and give exact data related to the duration and magnitude of seismic events. Accelerometers are another sort of sensor that quantifies the acceleration experienced by ground movement to identify rapid movement such as falling rocks. The measurements were taken using broadband seismometers which detect both high- and low-frequency seismic waves, making them perfect for capturing a full spectrum of ground vibrations that allow distinguishing magnitudes of rock falls. All of these sensor types serve a special role in monitoring the complete range of seismic activity associated with rock falls.

9.3.2 Data Acquisition and Analysis

Data acquisition and interpretation is the next step, once seismic sensors detect ground vibrations. This balance allows signals from the sensors transmitted to a centralized data processing unit—typically in real-time. These traces are then processed with signal processing techniques to filter seismic for noise, leaving only the interesting events. After the arrival of ground shaking at each station, key parameters such as amplitude, frequency, and duration of the seismic waves are analysed to estimate the characteristics of the event. Viewing from the perspective of rock fall, what we need to know must include: where does the event come from, how large is the falling rock mass, and what are the mechanics controlling that fall. Algorithms, machine learning Rock falls seismic data detection Automation Speed Accuracy.

9.3.3 Seismic Waves Detection

The technique of seismic monitoring has been used in many rock fall hazard studies, producing informative results and providing early warning in some cases. An interesting case study is from Yosemite National Park, where seismic sensors detect if some rocks were beginning to fall from the high granite cliffs in the park. The information gathered from the data provided clues about how often and why rocks fall there, allowing updates to be made in rockslide warnings. In Norway, researchers used seismic sensors to keep tabs on rock falls from precarious fjords. Compared to similar methods, he says the use of a laser scanner and images along with seismic monitoring offers a more holistic approach to detecting and mitigating risk caused by rock fall. This indicates clearly that seismological monitoring of rock falls can be an effective tool to reduce the loss of lives and infrastructure.

9.4 TECHNIQUES FOR INFRASOUND MONITORING

Infrasound Monitoring: So-called infrasounds are sound waves at frequencies below the range of human hearing, typically below 20 Hz. Since low-frequency sounds can travel so far through the air, it should come as no surprise that they are often created by natural events (such as volcanic eruptions or avalanches and rock falls). The basic idea of infrasound monitoring is that when a rock fall occurs, there may be collisions between large rocks hitting the ground that will release an amount of energy and with sufficient force produce infrasound waves capable of propagating through the atmosphere. In theory, these waves could be detected by infrasound sensors, allowing both for the monitoring of rock fall events and immediate determination of their location. Infrasound signatures are attenuated less by obstacles like terrain or meteorological influences, making them useful for detecting activity in remote and inhospitable locations.

9.4.1 Types of Infrasound Sensors

Infrasound sensors are also sensitive instruments that pick up low-frequency sound waves. Microbarometer: This type measures very slight changes in atmospheric pressure, due to the infrasound waves. The instruments, called microbarometers, are extremely sensitive to pressure, and can feel very small variations in air pressure caused by events at great distances away such as rock falls. Microphone array: A further type of sensor used for monitoring infrasound is the microphone array, composed of multiple microphones positioned in a geometrical shape to detect the direction and power of incoming infrasound waves. Researchers can use data derived from these arrays to track the origin and intensity of an event that created a similar infrasound. With both types of sensors acoustically presiding over the rock fall detections, they each bring information on the acoustic signals that are created surrounding these events.

9.4.2 Data Acquisition and Analysis

Acquisition and analysis of infrasound data start with capturing the signals from the sensors with which they are detected (infrasound waves), and their subsequent transmission to a processing centre of these signals. This pressure changes over time data from infrasound sensors. We then employ Fourier transform and waveform plotting to process the data, with greater use of more sophisticated signal processing that filters out all wind drift noise and only leaves signature infrasound signals generated by rock fall. The signatures of the rock fall, including its size and location, are obtained by analysing key parameters like amplitude, frequency, and duration of the infrasound waves. Algorithms are furthermore employed to serve as middleware for cross-checking infrasound data with seismic data so that the event is better defined, enhancing detection confidence and location accuracy.

9.4.3 Rock Fall Infrasound Detection Case Studies

Various empirical case studies have shown that the use of infrasound monitoring increases the capacity to detect rock falls in comparison with seismic sensors only, especially when these acoustic methods complement one another. Another key illu tration is the deployment of infrasound monitoring in the region of the Matterho Mountain in the Swiss Alps, where infrasound sensors were placed to search for an identify rock falls. However, the data collected allowed authorities to be able to develc early warning systems for hikers and tourists. It is also working on a case study i Canada with rock falls in the Canadian Rockies and using sensors to monitor infra sound waves. Here, the application of infrasound permitted the identification of loca tions and real-time observations in remote parts, affecting disaster control over nearb infrastructure and transportation routes. These case studies demonstrate the capabili of infrasound technology to improve rock fall detection and risk mitigation strategies

9.5 SEISMIC-INFRASOUND INVERSE MODELLING

Seismic and infrasound measurements offer complementary advantages in monitorin rock falls and together create a synergy that improves detection even further. Seismi sensors record any rocking of the Earth and vibrations that result, key to measurin rock fall impacts; and infrasound sensors pick up low-frequency sound waves gene ated by an event, which can be detected over long distances. Together, these two dat streams yield a more complete picture of rock fall events. Seismic sensors can detect if rock hit the ground and estimate the size of the rock fall, whereas infrasound can locat where in the atmosphere these bolides happened. These provide additional informatio and can be further analysed, or used in real-time monitoring in remote locations wher a domestic solution might not suffice.

$$\mathbf{D} = \mathbf{A} \cdot \mathbf{x} + \mathbf{n}$$

$$\mathbf{A} = \begin{bmatrix} \int_0^T G_1(t,\tau)\cdot\psi_1(\tau)d\tau & \int_0^T G_1(t,\tau)\cdot\psi_2(\tau)d\tau & \cdots & \int_0^T G_1(t,\tau)\cdot\psi_N(\tau) \\ \int_0^T G_2(t,\tau)\cdot\psi_1(\tau)d\tau & \int_0^T G_2(t,\tau)\cdot\psi_2(\tau)d\tau & \cdots & \int_0^T G_2(t,\tau)\cdot\psi_N(\tau) \\ \vdots & \vdots & \ddots & \vdots \\ \int_0^T G_M(t,\tau)\cdot\psi_1(\tau)d\tau & \int_0^T G_M(t,\tau)\cdot\psi_2(\tau)d\tau & \cdots & \int_0^T G_M(t,\tau)\cdot\psi_N(\tau) \end{bmatrix}$$

$$\mathbf{D} = \begin{bmatrix} d_1 \\ d_2 \\ \vdots \\ d_M \end{bmatrix}, \quad \mathbf{x} = \begin{bmatrix} x_1 \\ x_2 \\ \vdots \\ x_N \end{bmatrix}, \quad \mathbf{n} = \begin{bmatrix} n_1 \\ n_2 \\ \vdots \\ n_M \end{bmatrix}$$

$$G_i(t,\tau) = \frac{\partial}{\partial \tau}\left(H_i(t)\cdot \delta(t-\tau)\right) + \int_{-\infty}^{\infty} \frac{\partial H_i(t)}{\partial t} \cdot R_i(t-\tau)\,dt$$

$$H_i(t) = \sum_{j=1}^{K} C_{ij} \cdot e^{-\frac{(t-\mu_j)^2}{2\sigma_j^2}}$$

$$R_i(t-\tau) = \sum_{k=1}^{L} B_{ik} \cdot e^{-\frac{(t-\tau-\nu_k)^2}{2\sigma_k^2}}$$

Symbols Explanation:

- **D** is the data vector containing observed seismic and infrasound signals.
- **A** is the forward modelling operator matrix that relates model parameters to the observed data.
- **x** is the vector of model parameters to be estimated.
- **n** is the noise vector representing uncertainties in the data.
- $G_i(t,\tau)$ is the Green's function representing the response of the medium to an impulse at time τ.
- $H_i(t)$ is the source term for seismic signal associated with the ith sensor.
- $R_i(t-\tau)$ is the response function characterizing the medium's effect on signal propagation.
- C_{ij} is the coefficient for the jth component of the source model for the ith measurement.
- K is the total number of source components.
- L is the total number of response terms.
- μ_j is the mean time for the jth source component.
- σ_j is the standard deviation for the spread of the jth source component.
- B_{ik} is the coefficient for the kth response term for the ith sensor.
- ν_k is the time offset for the kth response term.
- ρ_k is the standard deviation for the spread of the kth response term.
- T is the total observation time.

9.5.1 Combining Data Sets

Combining seismic and infrasound data is challenging due to the variability of these measurements. The most typical one is time synchronization where both sensor types are merged by event time. This means that researchers can compare seismic and infrasound signals, pinpointing how ground vibrations relate to sound waves moving through the atmosphere. Geospatial triangulation is another technique that evaluates location data from both seismic as well as infrasound sensors to give a rock fall precise origin. Data assimilation is now more advanced with the use of machine learning algorithms to sift through both seismic and infrasound data looking for patterns that correspond to different types of rock falls. Since combining scores from these techniques provides a more accurate reading for the monitoring system, fewer false positives will be generated.

9.5.2 Improved Detection and Location Accuracy

Combining seismic and infrasound data greatly increases rock fall detection and location accuracy. Seismic sensors can fail to detect airborne events, and infrasound sensors may not provide a precise representation of the ground impact. However, when combined, the flaws of one system is compensated by the strengths of the other. The triple method slides in and catches the whole movement both going through and striking rocks into the ground. Moreover, by using data from many seismic and infrasound stations, it is possible to map the event with a higher resolution—we can measure the rock fall even more accurately in its location, size, and dynamics. This improved accuracy has high importance in early warning systems for the timely response towards anticipated threats before any damages become destructive in geotechnical engineering.

9.6 DEVICE DESIGN AND PRODUCTION

Given the harsh environmental conditions in which these sensors will be operating, they must be designed to survive without damage through wild weather and other extreme outdoor events. What has to be considered for deploying seismic and infrasound devices monitoring rock falls is not just the environment but also their respective sensitivity and durability. Characteristics: Seismic devices must be able to detect minute ground vibrations under all environmental conditions (variation of temperatures, nourishment levels, etc.). Likewise, sensor nodes for infrasound detection must be able to detect low-frequency sound waves with high granularity in the field among rocky areas where rock falls traditionally occur. Power efficiency is also key since many monitoring stations are located in remote areas where it is not as easy to perform regular maintenance. Seismic and infrasound devices serve as key components of a comprehensive monitoring network, due to the capability of transferring real-time data via integration with data acquisition systems and wireless communication modules.

9.6.1 Calibrating and Placing Sensors

The accurate detection and location of rock falls are highly dependent on the correct positioning of seismic and infrasound sensors. Ground-based seismic sensors are usually embedded in the Earth, whereas infrasound sensors are located on stable platforms to better capture low-frequency atmospheric pressure waves. To be accurate, those sensors would have to be spaced out evenly across a large area—the way seismometers are now located in places where quakes are expected—forming kind of a wireless network that can triangulate the location of rock falls. It may be a lot of boring work but calibration is also necessary for these sensors to collect accurate data. Before we started monitoring the slope, we needed to calibrate the devices—i.e. set their sensitivity relative to the expected signals produced by rock falls and account for the background noise; also, all sensors need to be time-synchronized in order of data analysis in real-time.

9.6.2 Real-Time Monitoring Systems

Early Detection and Mitigation: Systems for continuous real-time monitoring of seismic and infrasound devices alert in near real-time potential rock fall events, providing valuable seconds also important to early warning and mitigation strategies. Most of the systems are comprised of a seismic and infrasound monitoring network tied to a central processing unit that takes in, processes, and outputs data on real-time. The ability for data to be stored and analysed on cloud-based platforms provides geotechnical engineers and safety personnel access to live updates from anywhere. These systems are also starting to incorporate machine learning algorithms, which can autonomously identify strange readings in the seismic and infrasound data (decreasing presumably false alarms). In other words, the new functionality provides a faster environmental response, enabling an expedited decision for where and how to mitigate rock falls if property or people are at risk.

9.6.3 Maintenance and Calibration Procedures

The longevity and accuracy of the seismic and infrasound sensors necessitate proper maintenance and calibration. Sensors situated in remote locations and exposed to tough environmental conditions must be checked for spoilage at regular intervals. All of these instruments must be calibrated daily to tune their sensitivities to, for example, seasonal variations that can influence the behaviour of seismic or infrasound waves. Remote calibration systems make it possible for technicians to configure sensor settings from other locations without the need to visit each hardware means. Cleaning, testing battery performance, and recalibrating for accuracy need to be performed at times so that the monitoring system is operational over the course of months or even years thus supplying data continuously and reliably enough to detect rock fall.

9.7 DATA ANALYSIS AND INTERPRETATION

Seismic and infrasound monitoring data analysis refers to a wide range of algorithms and techniques that are used for discovering knowledge from the raw sensor denotes Fourier transform changes the time-domain data to frequency information for determining at which frequencies and amplitudes seismic and infrasound signals are peaking. Application of wavelet transform in non-stationary signal analysis—this is critical for detecting transient rock fall events. Seismic and infrasound algorithms classify rock fall events, and support vector machines (SVMs) classification patterns could be used in predicting rock falls with machine learning techniques, such as SVM and neural networks. Using historical data, these algorithms can spot patterns that always appear before rock falls to make detection and prediction more effective. Statistical methods and regression analysis are additionally employed to relate seismic and infrasound data to rock fall characteristics.

9.7.1 Signal Processing Methods

Signal processing techniques are of prime importance to sift out and analyse the information obtained from seismic as well as infrasound sensors. This is achieved using filtering techniques, including low-pass and high-pass filters, to erase noise and extricate the useful signals from background stimulations. One possible strategy consists of a typical amplitude modulation signal processing; this envelop detection helps to point out the intensity and duration of the presence of those events. A set of cross-correlation methods is used to compare seismic and infrasound signals so that they can be time-aligned with each other to study their temporal relationship, which is the key to improving detection capability. The advanced spectral analysis enables a detailed investigation of the frequency content of the signals, providing additional information on rock fall characteristics and facilitating discrimination between diverse types of seismic and infrasound events.

9.7.2 Seismic and Infrasound Data Analysis

We combine the processed signals to support a broader range of rock fall event interpretations from seismic and infrasound data. Seismic data essentially materializes the ground vibrations and reveals the impact forces, which are used to analyse how large a rock fall is, its location, etc. On the other hand, information from infrasound data gives us a measure of the airborne acoustic signals that allows an estimation of both the distance and orientation along which rocks fall. Multiplying these sources of data, it is possible to cross-measure this information and obtain higher precision in the interpretation made by researchers. Boulders of rock create a large seismic signal with an accompanying matching infrasound signal by combining the analysis of both datasets n great detail; the rock fall's dynamics can be better understood such as its size and peed on impact.

).7.3 Example: A Study on Rock Fall

\s an example of data analysis and interpretation, we can show a rock fall event moniored in Zion National Park, a case study that implements such practices. For the event, ve deployed a combination of seismic and infrasound sensors. The seismic analysis howed strong ground vibration with the plane of rock form clearly identified, demontrating an amplitude increase during impact caused by the second rock fall. At the same ime, infrasound sensors picked up a low-frequency acoustic wave that had travelled a ong way. Together, the datasets enabled the researchers to precisely calculate where the ock fall occurred, how big it was, and how much force it would have exerted. This is ow, with machine learning algorithms applied on their collected data, they have creited a predictive model to better predict early warning for future events. This case study hows how the integration of seismic and infrasound data enhances real-world applicaions, demonstrating that advanced data analysis methods can drastically improve rock all detection and protect human lives.

9.8 APPLICATIONS IN GEOTECHNICAL ENGINEERING

One of the most dangerous and expensive obstacles facing those in mining are rock falls. Using seismic and infrasound monitoring systems offers a very low cost permanent 24/7 monitoring solution for rock fall events providing continuous site safety and operational efficiency. Ground vibrations due to falling rocks are a strong indicator of unstable ground in the mining tunnels or open pits, which can be detected by using seismic sensors. Infrasound sensors can detect this feature by monitoring low-frequency acoustic signals caused by rock movement and are especially efficient for detecting the transition of rock falls from nearby to distal. These systems together with mine monitoring equipment allow the mining operator to monitor a dangerous area in real-time and take immediate action—for instance, evacuating personnel or changing mining strategy to prevent potential accident damage. By integrating these technologies, worker protection is increased as well as operational interruptions caused by rock falls are reduced.

9.8.1 Determining Stability on Construction Projects

The right determination of the geological status is indispensable for making arrangements to construct buildings in regions that are vulnerable to rock falls and landslides.

Monitoring of seismic and infrasound data will provide important information for monitoring the combined stability of slopes and excavation sites. Detection of more subtle ground vibrations that may signal shifts in the geographical structure is facilitated through seismic sensors, and tracking acoustic signals, indicating changes in the stress and movement within rock masses are possible through infrasound sensors. Engineers can use this data to determine the risk of future rock falls and landslides so that reinforcement and stabilization measures can be made. This leaves a positive outlook on the stability of projects in that they contribute to less risk and improve infrastructure through safer outcomes.

9.8.2 Prevention Methods and Early Warning Systems

Early warning systems using seismic and infrasound monitoring systems are a productive approach to reduce the hazards of rock falls. Such systems are meant to spot the first signs of a rock fall, for example, vibrations or acoustic signals, and issue warnings well in advance. By setting levels (thresholds) on seismic or infrasound data which represent decided indications of rock fall activity, these systems can initiate automated responses like alert notifications to safety personnel, activation of evacuation procedures, or changes in machinery operation. Subsequently, preventive measures such as slope support work or rock netting, variations of excavation schemes and designs can be implemented in view of the real-time data and early warnings generated by these monitoring systems. A combination of advanced analytics and predictive modelling raises early warning systems to a new level, resulting in the ability for early intervention that ultimately saves both human lives and technological assets alike.

9.9 LIMITATIONS AND CHALLENGES

There are many technical and environmental challenges for seismic and infrasound monitoring system deployment in geotechnical engineering. The technical complexities include sensor calibration and reliability at submicrometre level for real-time measurements on different surfaces. However, the detection of such low-amplitude vibrations can be difficult—even when using seismic sensors—as background seismic activity in other frequencies can cause interference with measurements. Sensors designed to detect infrasound are also confounded by several extraneous mechanisms, such as ambient winds and machinery in operation, which can mimic rock fall signals. Environmental challenges mean extreme conditions, which are common to the remotest and very rough terrain. They are subject to harsh environmental conditions like temperature extremes, moisture or dust exposure, and potential physical impacts.

9.9.1 Seismic and Infrasound Devices Drawbacks

Seismic and infrasound devices, despite being powerful tools, have their inherited limitations, which may affect the efficiency of rock fall detection. Seismic sensors can provide some information, but "they might not get the full picture of a rock fall" especially when ground vibrations are minimal, she said. They also need to be placed and aligned properly to accurately detect any signals, and can be sensitive to environmental noise as well as local geological conditions which can interfere with readings. Infrasonic sensor arrays are available commercially, although the sensitivity of these systems to low-frequency signals when influenced by ambient noise limits their range. Atmospheric conditions, such as temperature and wind, can affect the accuracy of infrasound measurements. Both types of sensors also demand real estate in the form of a dense network, which is further expensive and challenging to deploy and keep running.

9.9.2 Conclusion and Future Works

Last Mile and Way Forward: Solutions and future directions to overcome the challenges and limitations of seismic-infrasound monitoring developments in sensor technology are concentrating on sensitivity and reliability improvements of seismic as well as infrasound devices. Applications will range from improved sensors (with better noise filtering and calibrations) The combination of multiple monitoring techniques (seismic, infrasound, and optical sensors) will give a broader perspective on rock fall events to increase the ability in detecting them. You can use wireless and solar powered systems that work across remote locations or isolated power sources, which means easier maintenance and data transmission. The interpretation of data is also a growing area, utilizing machine learning and data analytics to improve the ability to differentiate rock fall signals from background noise. Opportunities will be sought through the development of cheaper and larger monitoring systems to make these suitable for use in geotechnical engineering.

9.10 CASE STUDIES AND REAL-WORLD IMPLEMENTATIONS

Seismic and infrasound monitoring systems are now being successfully deployed in the field to mitigate rock fall hazards, as multiple case histories also attest to. For decades, Yosemite National Park has used a network of seismic and infrasound sensors to keep an eye out for rock falls among known hot spots. Many other notable rock falls were detected including a near-miss event, where timely evacuations and protective measures were taken. Likewise, seismic monitoring systems have been used in Andean mines to record ground vibrations and predict the occurrence of falling rocks before they fall into mining work areas. Incorporation of infrasound sensors in these environments added an unprecedented capability to monitor rock falls, adding significant data sets that were used for safety and production decisions. These applications demonstrate the utility of combining seismic and infrasound technologies to create a robust and as well, inherent monitoring solution in many different settings.

Practical Scientific Field Deployments: Modern seismic and infrasound monitoring systems have been deployed for field experimentation, giving insights into their potential applications and practical usage. Calibration and sensor maintenance: This forms one of the fundamental lessons, that for long-term operation and reliable data, it is imperative to continuously verify the calibration of our sensors. Many sensors needed regular recalibration, which included temperature changes and contamination from dust in some instances. Sparse sensor networks sometimes missed smaller rock fall events or did not provide enough data points to accurately locate the event, indicating the importance of network density for effective monitoring. Also factored in was integration with existing safety protocols; the systems that could work in concert with evacuation and response plans were more likely to be useful as actual preventatives. In addition, data interpretation challenges led to the necessity of new methodologies such as advanced algorithms and machine learning methods for better identification of rock fall signals among background noise and other radiation sources.

9.10.1 Comparative Analysis of Monitoring Types

The strengths and weaknesses of different strategies are shown by comparative analysis of different monitoring methods. For instance, seismic-only systems are good at picking up ground vibrations but might miss an airborne signal of rock falls; infrasound-only systems detect the same lower-frequency acoustic signals (50 Hz), but provide little information about ground impact. This has been more effective, illuminating the full nature of rock fall events, according to volume. Optical and radar technologies have been used to supplement seismic and infrasound sensors when data on the movement and displacement of rock is required, for example. This multi-model approach can lead to better detection and location of rock falls, giving a higher performance. On the other hand, it needs more intricate data integrations as well as costs. The best monitoring approach for a site depends on the site's conditional budget and the level of accuracy required to measure that identifies a tailored approach is needed to respond to each different monitoring scenario.

9.11 FUTURE STUDIES AND INNOVATIONS

Rock fall monitoring can be further improved with new trends in sensor technology. Through this kind of sensor technology, advancements in materials are benefiting and making durable sensors to work even with extreme environmental conditions.

Miniaturization things are rounded, and sensors have greater reach with less intrusiveness. Increased signal precision is something else revolutionized by the new age, as ever more refined sensors are being used to boost detection with measurements that were simply inconceivable in the past years. Moreover, wireless and low-power technologies are evolving which permit sensors to work without demanding them to be hard-wired or necessitating frequent battery replacements in the most critical regions. Integration of multi-modal sensors, for example, fusing seismic and infrasound sensors with optical cameras yields a big-picture monitoring approach with a much richer data set and higher possibility of detection rates than the existing one when sensor redundancy is achieved. These technology improvements should increase overall reliability, precision, and ease of use in monitoring systems to help manage the danger from rock fall hazards.

9.11.1 Integration with Other Monitoring Systems Engage the Rest of Your Monitoring Ecosystem

Using their technologies in concert with other methods to learn, NIST points out that all three companies appear to be finding a niche offering what it calls "the complete rock fall detection solution." GIS (Geographic Information Systems) and other geospatial technologies can be used to overlay monitoring data on topographic and geological maps, giving a context how rock fall events are interpreted. Drones equipped with camera/lidar having high resolution can supplement sensor information providing live visual and topographical data, thus improving situational awareness. Furthermore, the synergistic installation of monitoring systems with IoT (Internet of Things) platforms enables uninterrupted data harvesting, transmitting, and analysing, which is useful for keeping things up-to-date and remotely operable. To improve the accuracy of rock fall detection and location, data from different sensors and monitoring systems are fused using data fusion techniques. This comprehensive way of working provides a more efficient system to help decision-makers monitor and consider the risks posed by rock fall.

9.11.2 Real-Time Rock Fall Detection Using Predictive Analytics and Machine Learning

Predictive analytics and machine learning are revolutionizing the detection of rock falls, making risk assessment more proactive and precise. Historical seismic and infrasound sensor data can be analysed by machine learning algorithms—that is, to recognize patterns in the data over time, then predict future rock fall events based on current conditions. For predictive mode, techniques like "neural networks, SVMs, and decision trees" are used. With thousands of models for predictive analytics, the technology can sift through real-time data and perhaps provide insights early enough to sound a warning or even automatically trigger actions that will prevent accidents and some of the damage. The algorithms for anomaly detection are used to find abnormal pattern in sensor data, which would suggest an upcoming danger of falling rocks. Big data analytics can be combined with real-time monitoring systems to allow for complicated processing of large volumes of historical and live data to make accurate predictions. As machine learning technologies level up, they will provide very accurate and timely rock fall detection for confident decision-making surrounding the risks of rock falls.

9.12 CONCLUSION

Therefore, predicting rock falls correctly is essential when it comes to the safety and protection of infrastructure in the field of geotechnical engineering. Approaches to monitoring rock falls are urgently required because they represent important hazards induced due to both natural and human factors. The coupling of seismic and infrasonic technology provides an efficient means for detecting and localizing these events in real-time. Both seismic, which can detect ground vibrations, and infrasound monitoring, capable of capturing low-frequency sound waves, offer different views of what happens when rocks fall. When we put data from both systems together, we greatly improve the accuracy and resiliency of detection systems for a faster, more sophisticated response to threats. The design and implementation of these systems call for the careful consideration of a variety of factors, including sensor placement, calibration, and ongoing maintenance. In addition, it is the advanced algorithms for data analysis that are crucial to translating seismic and infrasound signals into sound decisions for geotechnical engineers. The case studies clearly show the direct applications here, especially in industrial fields such as mining and construction sites where early warning systems could save accidents and cut costs on unforeseen damage. There remain numerous challenges to be overcome, including environmental disturbances, and the constraints of the device itself; however, we expect new sensor technology and machine learning techniques will continue to revolutionize this field. In the future, gauging is marked for seismic and infrasound systems of rock fall dynamics—illumined by predictive analytics secured from data acquired by a consistent histogram will characterize most interpretations of rock fall in geotechnical engineering fields with a triumphant volume addressed accordingly and safety-centric requirements will enforce this arrangement.

REFERENCES

Amitrano, D., Grasso, J. R., & Senfaute, G. (2005). Seismic precursory patterns before a cliff collapse and critical point phenomena. *Geophysical Research Letters*, *32*(8), 1–5.

Collins, B. D., & Znidarcic, D. (2004). Stability analyses of rainfall induced landslides. *Journal of Geotechnical and Geoenvironmental Engineering*, *130*(4), 362–372.

Deparis, J., Jongmans, D., Cotton, F., Baillet, L., Thouvenot, F., & Hantz, D. (2008). Analysis of rock-fall and rock-fall avalanche seismograms in the French Alps. *Bulletin of the Seismological Society of America*, *98*(4), 1781–1796.

Dietze, M., Mohadjer, S., Turowski, J. M., Ehlers, T. A., & Hovius, N. (2017). Seismic monitoring of small alpine rockfalls–validity, precision and limitations. *Earth Surface Dynamics*, *5*(4), 653–668.

Feng, Z. Y., & Zhuang, R. C. (2021). Characteristics of seismic and acoustic signals of rock falls: An experimental study. *Landslides*, *18*(11), 3695–3706.

McNutt, S. R., Thompson, G., Johnson, J., De Angelis, S., & Fee, D. (2015). Seismic and infrasonic monitoring. In Leighton M. Watson, Alexandra M. Iezzi, Liam Toney, Sean P. Maher, David Fee, Kathleen McKee, Hugo D. Ortiz, Robin S. Matoza, Julia E. Gestrich, Jordan W. Bishop, Alex J. C. Witsil, Jacob F. Anderson, Jeffrey B. Johnson (Eds.), *The Encyclopaedia of Volcanoes* (pp. 1071–1099). Academic Press.

Schmandt, B., Aster, R. C., Scherler, D., Tsai, V. C., & Karlstrom, K. (2013). Multiple fluvial processes detected by riverside seismic and infrasound monitoring of a controlled flood in the Grand Canyon. *Geophysical Research Letters*, *40*(18), 4858–4863.

Olivieri, G., Vezzosi, S., Farina, P., & Meier, L. (2020). On the use of acoustic records for the automatic detection and early warning of rockfalls. In *Proceedings of the 2020 International Symposium on Slope Stability in Open Pit Mining and Civil Engineering* (pp. 1193–1202). Perth: Australian Centre for Geomechanics.

Walter, M., Arnhardt, C., & Joswig, M. (2012). Seismic monitoring of rockfalls, slide quakes, and fissure development at the Super-Sauze mudslide, French Alps. *Engineering Geology*, *128*, 12–22.

Xie, T., Zhang, C. C., Shi, B., Wang, Z., Zhang, S. S., & Yin, J. (2023). Seismic monitoring of rockfalls using distributed acoustic sensing. *Engineering Geology*, *325*, 107285.

Yan, Y., Cui, Y., Guo, J., Hu, S., Wang, Z., & Yin, S. (2020). Landslide reconstruction using seismic signal characteristics and numerical simulations: Case study of the 2017 "6.24" Xinmo landslide. *Engineering Geology*, *270*, 105582.

Zimmer, V. L., & Sitar, N. (2015). Detection and location of rock falls using seismic and infrasound sensors. *Engineering Geology*, *193*, 49–60.

10 Advancing Slope Immovability in Undeveloped Cast Mines via Machine Learning-Based IoT Structure

10.1 INTRODUCTION

Slope stability in mining is a critical aspect that has a direct bearing on both the safet and productivity of mining activities. With changes to the mining industry over time the process of managing and maintaining stable slopes has become more difficult, especially in cast mines where traditional methods often fail. These mines are unique i mineralogy and extraction techniques, making them prone to instability from differences in rock properties and mining conditions (Das et al., 2017).

Slopes in open-pit mines are exposed to the constant impact of dynamic force due to mining operations as well as natural causes. The inherent variability in roc behaviour, coupled with the enormity of operations, restricts monitoring and predictio of slope stability by conventional means only. The deeper the mining operations nee to reach and the more complex its geology, and the chances of slope failure increase thereby leading to risks that can be a threat for workers as well as hamper the extractio of minerals (Sharma et al., 2021).

DOI: 10.1201/9781003537625-1

In view of these challenges, the inclusion of Internet of Things (IoT) technology and machine learning (ML) is a groundbreaking solution. Through IoT, it becomes possible to deploy a network of sensors and devices that are continually monitoring several parameters that affect slope stability, such as soil moisture, strain, and displacement (Ghorbani et al., 2020). Webcams offer a live snapshot of slope conditions, with data that are collected and analysed in real-time, giving Creekside more up-to-date information to make timely assessments of slope stability.

Given the volume of information that IoT can generate, ML becomes an ideal pairing with IoT systems, as it can sift through mountains of data to determine any patterns in resource utilization and predict where any failures might occur. This predictive model helps in understanding the causes contributing to slope instability and aids in predicting the future orientation of the slopes. This prediction ability improves decision-making while enhancing system safety and operational efficiency (Chen et al., 2020).

This resistance often disguises the true potential of what can be achieved when we integrate IoT and ML into slope stability management, contributing to a more robust and proactive monitoring system. Advanced predictive analytics enable mining companies to shift resource management from reactive to proactive by combining real-time data acquisition with validated, detailed forecasts, and coaching. This shift leads to the early recognition of potential problems and proper intervention, thereby minimizing the risks involved with disastrous failures in slopes.

The purpose of using IoT and ML in slope stability monitoring is also to increase the accuracy and reliability of stability assessments. Typically, traditional methodologies depend on limited data and manual observations that are not well-suited to handle the dynamism of slopes. Conversely, an IoT system provides real-time monitoring and data changes on the spot, while ML models can be adjusted based on new information, delivering better predictions and advice.

The deployment of IoT and ML, in addition to effectively tackling complex technical issues related to slope stability, also leads to improvements in overall productivity. Through automation of data capture and analysis, these technologies eliminate the need for manual inspection and intervention, enabling more cost-effective allocation of resources and reducing downtime. In turn, this operational efficiency leads to cost reduction and productivity improvement for mining companies (Ghorbani et al., 2020).

Furthermore, the use of these technologies allows for building extensive knowledge on slope stability by adding numerous sources and variable data. This whole-of-mine approach allows the Geo Database Management Centre (GDMC) to develop better management strategies and appropriate solutions intended specifically for individual mining sites. These data stream includes the ability to analyse and interpret complex readings from multiple sensors and sources, providing stability monitoring with a comprehensive view.

10.2 SLOPE STABILITY FUNDAMENTALS

Slope stability analysis is one of the most important ground tasks in geotechnical engineering, designed to evaluate and check if slopes are safe and stable across a number of possibilities such as that involved with mining operations. Slope stability analysis is, at its heart, the evaluation of competing forces that encourage movement versus those that counteract it. Central to these include identifying the possible causes for providing stability of a slope and recognizing what processes can lead to sliding (Table 10.1).

A key principle (one of many) is the comparison to shear stress vs. shear strength. Soil type and soil moisture content, as well as the presence of weak planes or joints that result in increased loading on summertime permafrost features, have been shown to cause more significant drainage failure (Zhang et al., 2019). The vertical pressure, however, changes due to the overlying weight and external loads, such as earthquakes or larger amounts of rainfall; whereas, redistribution is affected by the sheer force. The stability of slopes is assessed by comparing the two factors, indicating whether the slope will remain stable or fail.

The other principle is the analysis of the factor of safety (FS), which in this case refers to a ratio of resisting forces to driving forces. Slopes with a FS higher than one are considered stable, while slopes with a FS lower than one are at risk. This principle establishes an index for grading slope stability and an acceptable basis for determining engineering options.

10.2.1 Factors Controlling Slope Stability in Cast Mines

There are a number of controlling parameters that influence slope stability in cast mines, which, together or individually, will contribute to the global slope stability or instability. The geological and geotechnical properties of the rock or soil are among the most important influencers. The strength of the rock, soil cohesion, and the presence of natural fractures or faults that cut across slopes can all significantly affect slope stability (Singh & Sharma, 2021).

This is also important for the mining method itself. Disruption of the natural stress field and excavation of large volumes of material in open cast mines may cause appreciable changes in slope stability due to the creation of voids. If materials were to be removed from beneath a slope, it may lead to a reduction in supporting forces, causing the feature to collapse.

Environmental conditions like rain and infiltration of underground water can worsen slope instability. Pour water can also decrease the shear strength of soil and rock by increasing pore pressure, giving rise to additional saturation of a portion or the entire thickness. Extreme temperature changes, in conjunction with freeze-thaw cycles, can also contribute to ground movements and slope instability (Zheng & Tang, 2020).

TABLE 10.1 Fundamentals of slope stability

FIELD	*CONCEPT*	*DESCRIPTION*	*FACTORS AFFECTING STABILITY*	*MITIGATION MEASURES*
Basic principles	Equilibrium analysis	The equilibrium of forces acting on a slope is assessed to determine if it is stable or unstable. The concept of forces includes gravitational forces, shear forces, and normal forces.	• Weight of soil and rock • Cohesion of soil • Internal friction angle • Water content	• Regular monitoring of slopes • Implementing proper drainage systems to reduce water infiltration
Types of failure	Sliding	A common failure mode where a mass of soil or rock moves down a slope along a failure surface, which can be classified as rotational or translational sliding.	• Slope angle • Type of material • Presence of weak layers • Weather conditions	• Soil nailing • Retaining walls • Grading and flattening of slopes
Stability analysis methods	Limit equilibrium method (LEM)	A widely used method for analysing slope stability that evaluates the forces and moments acting on a potential failure mass to determine the factor of safety (FS) against sliding.	• Geometry of the slope • Material properties • Groundwater conditions	• Increasing the FS by modifying slope geometry • Improving soil properties (e.g., compaction)
Influence of water	Pour water pressure	The presence of water within soil pores can reduce effective stress, increasing the likelihood of slope failure. Pour water pressure is a critical factor in stability analysis.	• Heavy rainfall • Melting snow • Groundwater table fluctuations	• Drainage systems (e.g., ditches and pipes) • Pore pressure relief measures
Geotechnical investigation	Soil testing and characterization	Conducting tests (e.g., shear strength tests and grain size analysis) to characterize the soil properties is crucial for understanding slope stability.	• Soil type and composition • Shear strength parameters • Compressibility and permeability of the soil	• Comprehensive geotechnical investigations prior to construction • Regular assessments and monitoring

10.2.2 Traditional Techniques for Monitoring and Management of Slope Stability

Over the years, traditional methods have been used to monitor and manage slope stability in mining operations. It usually may be referred to as the combination of field observations, manual measurements, and empirical calculations to analyse and control slope stability.

Inclinometer and piezometer readings are among the typical traditional methods. Inclinometers measure the displacement and deformation of the ground — revealing whether or not a slope is moving. Piezometers record the levels of groundwater and pore pressure in underground areas, which is important for understanding the role that water plays on slope stability (Zhang et al., 2019).

Visual inspections and manual surveys aim to assess stress in cracks, bulges, or shifts of earth. While they are remarkable at identifying visible concerns, they frequently do not have the depth and range required for timely caution signals to be raised and proactive administration.

Similarly, slope stability analysis also depends on empirical methods and stability charts developed through historical data and engineering judgement. Although these tools offer incredible value, they neither fully recognize just how dynamic slope stability is nor the complexities in contemporary mining operations.

10.3 IoT IN GEOTECHNICAL ENGINEERING

Simply put, the IoT is a game changer where inanimate objects are connected to each other through their own network. It facilitates real-time data collection, communication, and analysis, which are the foundation for making well informed decisions and automation. From infrastructure health to soil conditions or slope stability assessments, IoT technology is applied in various fields of geotechnical engineering.

In the context of geotechnical engineering, IoT applications involve the deployment of several sensors and devices, which collect data on different environmental and structural parameters. The devices send this information to centralized systems where the data are collected and examined. Continuously monitoring parameters like soil moisture, ground movements, or structural condition allow engineers to have a full picture of the geotechnical phenomena and enable them to take actions at an early stage.

10.3.1 IoT Devices and Sensors for Geotechnical Monitoring

A variety of IoT sensors and devices are used to collect and send essential parameters in geotechnical monitoring in Figure 10.1. These include:

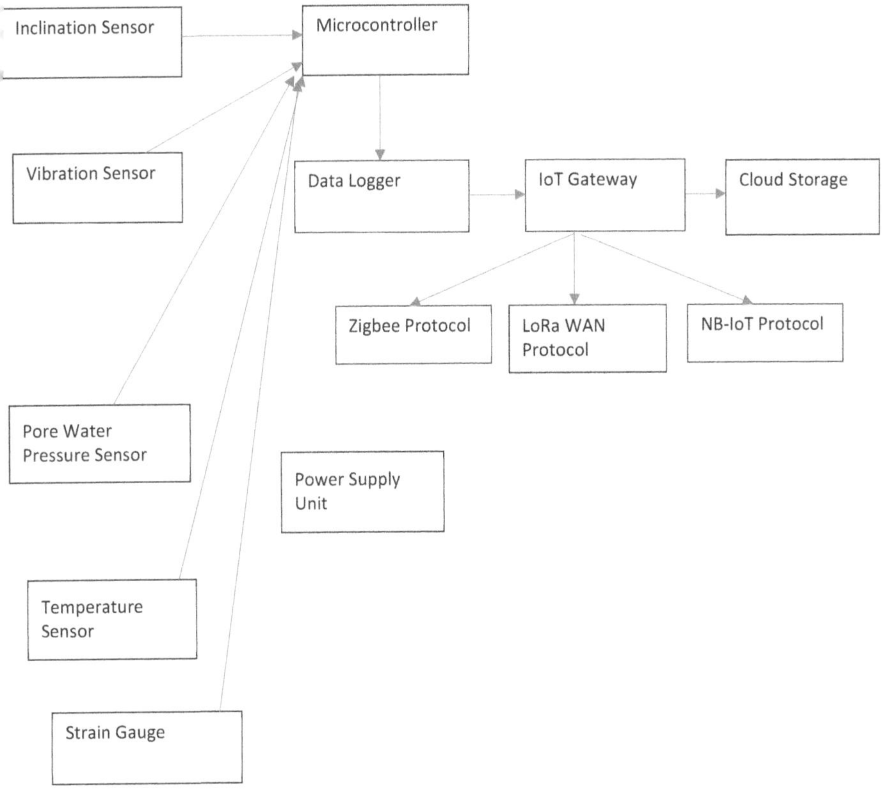

FIGURE 10.1 Internet of Things (IoT) devices and sensors for geotechnical monitoring.

Inclinometers: Inclinometer sensors are typically used to monitor ground deformation and slope movement, gaining valuable information about the stability of slopes and embankments. Piezometers measure pore water pressure; an important aspect to determine the head of groundwater in soil stability and slope behaviour.

Strain Gauges: Mounted on structures and rock formations, and these small sensors monitor stress/strain to support evaluation of looming structural failure.

Thermometers and Hygrometers: Tools of measuring temperature and humidity, respectively, to give track into environmental conditions.

Geophone/Accelerometer: Geophones are used to detect ground vibrations and seismic activities, which are important when measuring the impact of external forces to slope stability.

The devices use a drill and install process to function as a complete monitoring system that is able to collect key data for proper geotechnical management.

Over the past 5–10 years, we have seen the widespread application of IoT in mine sites, and this tech has potential to revolutionize how geotechnical data are collected and analysed. However, this also allows for continuous, real-time monitoring to help ensure businesses are running as predictably smooth. The information runs continuously thereby giving us opportunity to check potential issues before they turn into major ris aside being able to take action well in advance.

IoT systems improve data accuracy as well. Traditional approaches have relied c manual measurements and sporadic checks, call for time and effort, are error pron easy to overlook important shifts. Using IoT sensors is more accurate with the adva tages of being close to real-time, no human fallibility, and enhances reliability for st bility assessments.

Additionally, the combination of IoT technology can also reduce costs and impro operational performance. Doing away with the need for manual labour when collectir data also decreases downtime because of the time needed to do a full check man ally. Furthermore, more timely monitoring means real-time resource management ar a faster response time improving the overall speed of operations.

Yet, applying the technology of IoT in mining environments has problems as wel There are numerous challenges to clear, not the least of which is the tough enviro mental condition characteristic of mining sites. IoT devices — which need to run extreme temperatures, dust, and humidity that can affect their performance and dur bility — call for robust, specialized equipment.

It is facing similar data management and integration hurdles. The sheer volume c data that are generated by IoT sensors demands advanced data management systems fc efficient processing and analysis. To derive the most benefit from IoT technology, it important to secure that data integration — ensuring quality and security and combir ing this IoT-delivered geotechnical data with existing models in use at various organiz tions — remains an essential part of geoscience digitalization.

10.4 ML IN GEOGRAPHICAL ENGINEERING

ML: ML is the subfield of artificial intelligence (AI), which enable computer system to learn from data and make predictive decisions based on it. Rather than strictly fo lowing instructions as determined by the programmer, ML systems grow more effectiv over time because they learn to observe patterns and make data-driven decisions. It of particular interest in geotechnical engineering due to the complexity and organiza tion of data when dealing with datasets for which relationships may be non-linear an non-Gaussian around very high dimensions than what classic methods can handle.

ML is suitable in geotechnical engineering since it can gather and analys large-scale diverse data easily from soil testing, geophysical surveys, and sensor mea surements. ML algorithms will help engineers understand how geotechnical system behave and predict problems, and in that way, optimize the designs and interventions Such predictive capability is essential in the context of the mitigation of risks related t soil stability, infrastructure health, and environmental impacts, leading ML to becom a fundamental part of improving geotechnical engineering practices.

$$L = \sum_{i=1}^{N} (y_i - \hat{y}_i)^2 + \lambda \sum_{j=1}^{M} (\Theta_j^2)$$

$$\hat{y}_i = f(X_i, \Theta) = \sigma(W^T X_i + b)$$

$$\sigma(z) = \frac{1}{1 + e^{-z}}$$

$$\frac{\partial L}{\partial W_j} = -\frac{2}{N} \sum_{i=1}^{N} (y_i - \hat{y}_i) X_{ij} + 2\lambda W_j$$

$$\frac{\partial L}{\partial b} = -\frac{2}{N} \sum_{i=1}^{N} (y_i - \hat{y}_i)$$

$$W_j^{(t+1)} = W_j^{(t)} - \alpha \frac{\partial L}{\partial W_j}$$

$$b^{(t+1)} = b^{(t)} - \alpha \frac{\partial L}{\partial b}$$

$$\Theta = \begin{bmatrix} W^T, b \end{bmatrix}$$

$$X = \begin{bmatrix} X_1 \\ X_2 \\ \vdots \\ X_N \end{bmatrix}$$

$$Y = \begin{bmatrix} y_1 \\ y_2 \\ \vdots \\ y_N \end{bmatrix}$$

$$Y_{pred} = f(X, \Theta)$$

$$R^2 = 1 - \frac{\sum_{i=1}^{N} (y_i - \hat{y}_i)^2}{\sum_{i=1}^{N} (y_i - \bar{y})^2}$$

$$\text{AIC} = 2k - 2\log(L)$$

$$\text{BIC} = k\log(N) - 2\log(L)$$

Symbol Explanation:

- L is the loss function.
- N is the number of observations.
- y_i is the actual output for observation i.
- $\hat{y}_i$ is the predicted output for observation i.
- λ is the regularization parameter.
- M is the number of features.
- Θ is the parameter vector, including weights and bias.
- W is the weight vector.
- b is the bias term.
- $\sigma(z)$ is the activation function (sigmoid).
- α is the learning rate.
- X_i is the feature vector for observation i.
- R^2 is the coefficient of determination.
- AIC is the Akaike information criterion.
- BIC is the Bayesian information criterion.
- $\bar{y}$ is the mean of the actual outputs.

10.4.1 Types of ML Algorithms for Predictive Modelling

In geotechnical engineering, different kinds of ML algorithms are used to predict various aspects with one more workload and a complex dataset than the other. The training of a model on labelled data, where outcomes are known, is considered as supervised learning. Examples of supervised learning algorithms are listed below.

Regression Analysis: This is used in order to prepare prediction of continuous outcomes, e.g., soil settlement or slope displacement, etc. Some of the simple techniques that we covered earlier, such as linear regression and polynomial regression, belong to this category.

Classification Algorithms: This is Used when the outcome is categorical logistic regression, support vector machines (SVM), and decision trees and are typical algorithms that can be used to classify soil types or predict failure conditions, in just a simple example case.

Unsupervised Learning: This technique is for unlabelled data, which mean its goal to find hidden patterns or rosters. Key techniques include the following:

Clustering Algorithms: Like *k*-means or hierarchical clustering, which are used to group similar data points together. It is a technique that allows geotechnical data to be grouped into clusters based on properties such as soil category or stress condition.

Dimensionality Reduction: These techniques like PCA help in lowering the dimension of our complex data into fewer dimensions while preserving the important features. It is useful in visualising and breaking down complex data.

Reinforcement Learning: This includes training models for a series of decisions and reward them for getting a right result. This is less common in geotechnical engineering but has its applications.

10.4.2 Application of ML in Geotechnical Data Analysis

The use of ML in geotechnical data analysis has greatly developed the field by improving predictive accuracy and decision-making. A few of the most popular use cases are:

Predictive Maintenance: ML algorithms processing historical soil status conditions, infrastructure performance, and environmental variables to alert potential breakdowns or maintenance necessities. ML models can, for example, predict the occurrence of slope stability problems through observation and cross-referencing with historical incidents and real-time sensor readings.

Geotechnical Site Characterization: An ML model to interpret complex geological data such as borehole logs and seismic survey. ML models help to better and more quickly characterize site conditions by finding patterns and correlations in the data.

Risk Assessment and Management: It supports in measuring the risk related to multiple geotechnical problems like landslides, subsidence, etc. Based on current data, historical pattern predictive models can assess the probability of such events taking place and suggest mitigation strategies.

Design Parameter Optimization: By extracting performance data and simulations, ML-based algorithms can determine the optimal design parameters for foundations, slopes, and other geotechnical structures. This optimization results in more viable engineering solutions and at a lower cost.

Real-Time Monitoring: ML with IoT systems improve real-time monitoring rectories. ML models, for instance, can use sensor data to either detect anomalies or predict failures, effectively warning the operations team, so that they can take proactive measures.

10.5 INTEGRATION OF IoT AND ML FOR SLOPE STABILITY

Incorporating IoT systems to slope monitoring requires an architecture that allows full-spectrum data collection, real-time processing, and meaningful analysis. A design, in general, consists of a network of IoT sensors deployed across the entirety of a slope that captures pertinent parameters, influencing stability such as soil moisture, temperature, strain, and displacements (Hassan & Mahmoud, 2021).

10.5.1 Architecture of IoT-Based Slope Monitoring System

The architecture of an IoT-powered slope monitoring system mainly includes the following items:

Sensors and Devices: These are the main data-fetching devices. Inclinometers piezometers, strain gauges, and other sensors are placed to measure the different environmental and structural conditions. They acquire disparate data associated with slope stability (Gholami & Zarei, 2018).

Data Transmission: The obtained data are transmitted from the sensors to a centralized data store or cloud platform through communication protocols over wireless networks, cellular networks, or satellite links. Good data communication is required for the authenticity of the transmitted data and to keep updated in time (Li & Fang, 2021).

Data Storage and Management: Your data are stored in a database or cloud-based system where it can easily be accessed, managed, and used. This layer ensures that historical and real-time data are available for analysis and decision-making (Zhang & Wei, 2021).

Data Processing Unit: Having computational resources and algorithms that ingest raw sensor data. The data processing unit processes the clean, transformed, and aggregated form of raw data for initial preparation necessary before configuring advanced analysis (Madani et al., 2020).

User Interface and Dashboard: A User Interface (UI) or dashboard to display and make sense of the data. The web application includes graphs, charts, and alerts related to the slope conditions quickly seen by engineers and decision-makers (Kumar & Kaur, 2022).

10.5.2 Big Data Collection and Management from IoT Sensors

The continuous measurements of parameters influencing slope stability are collected using the IoT sensors. If the sensors are not accurate and sensitive enough and the data handling is not robust enough, then obviously an automatic herd system cannot work.

10.5.2.1 Data Collection and Management Overview

Sensor Calibration and Maintenance: Sensors must be calibrated and maintained on a routine basis in order to provide accurate readings. Routine confirmations are necessary as environmental influences and sensor drift can lead to inaccuracies in the measurements (Wang & Li, 2019).

Data Quality Assurance: Using certain techniques to validate the data collected n order to ensure its accuracy and consistency. Data purity is ensured by anomaly etection and error correction algorithms (Gholami & Zarei, 2018).

Data Integration: Merging data coming from a lot of sensors and sources to create one big dataset. This means combining data streams from different sensors, like nvironmental and structural ones, in a cohesive way to present a complete picture of he slope status (Madani et al., 2020).

Data Storage: Selecting the right storage that can be cloud-based or on-premise, o scale and store huge amounts of data. Long-term monitoring and analysis can be proided by data storage to ensure reliability, scalability, and security (Li & Fang, 2021).

10.5.3 Prediction of Slope Failures by ML Models

ML Models: ML models are used to process the data obtained from IoT sensors to orecast slope failures. These models help predict likely stability issues, which were leveloped via historical data and real-time inputs. Some of the essential features of ML n slope stability are:

Feature Extraction: Extracting relevant features from the sensor data related to slope stability. Issues may arise from different soil properties, stress levels, moisture content, and environmental conditions (Zhang & Wei, 2021).

Algorithm Selection: Deciding on the most suited ML algorithms, depending on he data at hand as well as objectives for prediction. The popular algorithms that are commonly used include decision trees, SVM, different neural networks, and ensemble nethods (Kumar & Kaur, 2022).

Model Training and Validation: This phase is essential for training the ML models with historical data and validating their performance by test datasets. This process nvolves fine-tuning model parameters and testing for accuracy to ensure that predicions are reliable (Hassan & Mahmoud, 2021).

Model Deployment: Embedding the trained ML models into the IoT system in order to provide real-time alerts and predictions. The models evaluate great quantities of incoming data to spot patterns from the past that suggest a potential slope failure (Li & Fang, 2021).

10.5.4 Real-Time Data Processing and Analysis

The key to the successful integration of IoT and ML in slope stability monitoring is real-time data processing and analysis. This involves:

Data Streaming and Analysis: Systems for processing streams of data from IoT sensors. Since data are processed in real-time, potential problems can be detected immediately during collection (Gholami & Zarei, 2018).

Predictive Analytics: Deploying ML models for real-time data analysis and generating predictions from trends. This entails continued application of algorithms to establish trends and patterns that can be indicative of likely slope movements in th future (Madani et al., 2020).

Alert Systems: The development of alert processes to inform engineers and oth involved stakeholders, such as decision-makers, is based on individualized data an model thresholds. For instance, an alert may be generated based on the exceedance thresholds or set parameters for certain timeline lengths (Wang & Li, 2019).

Visualization and Reporting: Provision of real-time visualization and reporti tools or dashboards to show the current conditions of the slope, the above-generate predictions, and alerts. Axial visualization is crucial in fast decision-making, as we as reporting on slope indices and history within the set period (Kumar & Kaur, 2022

10.6 CASE STUDIES AND APPLICATION

Some of the implementations or applications of IoT-based slope monitoring system have shown prove efficient for slope stability management and warning systems. A excellent example is the application of IoT techniques in the Mount Emei site, whe a system of sensors was installed to monitor slope movements and determination o soil aspects. The inclinometers and piezometers networked with environmental cen sors to collect data on a real-time basis and feed it to the central database for analysi The timely system helped in the monitoring of slope movements timely with advance actionable information and environmental factors ready to avert risks of sliding. Th application has succeeded in reducing erosion factors and the frequencies of landslide thus enhancing the safety of the local community. Another useful case study is th application of IoT in Kurobe Dam, where an elaborate system was installed to determin structural health and stability. Vibration censors and temperature gauges networke with displacement sensors were used to determine the stability of the dam and identif likely failure points. The timely real-time data coupled with predictive analytics wa efficient in data usage for targeted maintenance and structural safety.

10.6.1 ML Applications on Slope Stability Analysis

Over the years, interest and applications of ML have grown to analyse and predic slope stability, which has changed the way geotechnical engineers approach slope man agement. The most prominent application is to perform a data-driven analysis of th susceptibility to landslides in the Swiss Alps using ML algorithms. Based on histori cal landslide data, soil properties, and environmental constrains, predictive model were developed. Then, some powerful algorithms such as random forests and gradien boosting machine (GBM) were used for pattern identification and to predict landslid occurrences. These models have been effective to provide risk assessment guidance an support mitigation strategies.

For instance, ML was employed to manage and predict slope stability of an open-pit mining activity in Australia. The project used a mix of geotechnical data, sensor readings, and ML models to predict the real-time stability of the slope. Neural networks and SVMs were employed to uncover complex relationships among variables as well as to estimate slope failure. It also improved the forecasting accuracy and made predictive maintenance decisions much more robust, which in turn provided higher operational safety and productivity.

Using the characterization of two cases, IoT and ML were adapted to this task, and we learned a lot from it:

- **Why Data Quality:** In this digitized IT environment, high-quality data are necessary to make accurate analysis and predictions. In multiple projects, problems with inconsistent sensor calibration, data integration and measurement errors clearly indicated that more efforts were required to perform these steps timely and inclusive data quality control and validation procedures.
- Integrating IoT and ML systems to existing infrastructure provide good practices, and the reality is that there are still some technical challenges. To maintain a smooth data flow between sensors in vehicles, storage systems, and various analysis tools, it is necessary to arrange everything well and design complex system architecture.
- **Real-Time Processing:** The performance of real-time monitoring and analysis highly depends on system capabilities to quickly cope with tonnes of processed data. For instance, projects using efficient data processing frameworks and real-time analytics performance benefits were significant, which help in timely intervention.
- **Model Validation and Tuning:** This is to keep revalidating the models and adapting over a period of time. However, projects with feedback loops tuning into recommendations and rebalancing dependencies for changing data and conditions performed better in prediction.
- Receiving the most out of the data is necessary for proper implementation of these technologies, which is easy with user training and engagement. It is important, therefore, that engineers and those responsible for making decisions are well trained to interpret and translate the outputs of IoT and ML systems into actionable information.
- **Scalability and Flexibility:** The projects that initially approached their systems with motivations based on scalability and flexibility were better positioned to translate these uses into changing end-user needs as well as performing them in the future. This method keeps the system independent and functional as time.

10.7 CHALLENGES

Challenges to IoT and ML integration for slope stability monitoring Connectivity and Reliability – A major challenge is IoT networks in remote, harsh environments such as mining sites. Uninterrupted data transmission between the IoT devices and their central processing systems is hard to guarantee, especially when rough terrain, harsh weather conditions and infrastructure limitations predominate. Incomplete data collection due to network instability or inaccurate sensor readings can severely hinder the performance of a system.

The scalability of the system is another horrific technical issue. It handles high volumes of data in real-time: With more sensors and data points, it becomes harder to manage and process large amounts of data in real-time. However, IoT systems created to track small-scale monitoring will be hard-pressed in such complex environments like the mining sector where bottom of the pit data can take days just to understand.

The difficulties in building robust predictive models in ML mode. However, the geotechnical data are frequently so non-linear and dispersed that it makes the input-to-output function quirky. These complexities make designing models that can accurately capture these and make real-time predictions a considerable technical challenge. Furthermore, the robustness and adaptability of these models, so as to account for varying slope conditions over time, are essential to maintain their accuracy in order to predict shallow landslide hazard years into the future.

10.7.1 Eliminate Data Quality and Integration Issues

Data quality is one of the key problem in working with IoT and ML technologies. IoT sensors can suffer from calibration drift, environmental interference, and mechanical wear out that lead to bad or lost data. Even a single case of wrong data can destroy the model's accuracy and can return erroneous predictions making worse decision.

A big challenge in this is data integration. All the data from different IoT sensors need to be stitched together neatly into one single dataset. But this process can be problematic due to inconsistencies in data formats, synchronization problems, and missed data points. It becomes more complicated when incorporating geotechnical data points with climactic considerations such as localized temperature and rainfall. This often leads to less optimal ML models, which may in turn decrease the performance of said system, for the lack data itself.

10.7.2 Overcoming Challenges

However, in order to implement IoT and ML for monitoring the stability of slopes, a number of best practices and solutions need to be followed as follows:

Risk mitigation solutions include the use of mesh networks or satellite communication systems in remote and harsh environments, which increase data transmission reliability. These systems provide a mechanism for ongoing communication between IoT sensors and central servers, even where traditional network infrastructure is not available.

Design of Strong Data Processing Architectures: Configuring saleable cloud systems or leveraging edge computing to process data on premises can solve the challenge of managing huge datasets. Edge computing processing can be done in real-time at a data collection point, reducing latency and thus ensuring faster decision-making.

Advanced Sensor Calibration and Maintenance Protocols: Regular IoT sensor calibration and proactive maintenance schedules help maintain data quality. Self-calibrating sensors and automated diagnostics can improve sensor reliability, while periodic checks ensure data accuracy in the long term.

ML Model Validation and Adaptation: Ensemble learning techniques are shown to be particularly successful in coping with the inveterate non-linearities, uncertainties entrenched within geotechnical data. Embedding ensemble learning into more powerful hybrid models that employ multiple algorithms simultaneously would provide a solid strategy when the eventual goal is using such tools on site. This allows the models to be used over and over again with simple validation methods and allowing interesting flexibility in changing situations.

Data Cleaning and Preprocessing Tools: Making use of advanced data cleaning technique like outlier detection, interpolation, and smoothing data can be very crucial for improving the quality of your data before feeding it to ML pipeline. Standardized data formats and time-series synchronization protocols can be leveraged to solve data integration issues, providing homogeneous and full data in order for the system to analyse it with confidence.

Predictive Maintenance for IoT Systems: The addition of predictive maintenance algorithms to IoT devices can significantly cut down on downtime as well as sensor failures. These algorithms evaluate performance data of different sensors and detect potential failures long before they actually start, so that consequent repairing or replacing would be done in timely manner.

Training and Capability Building: It is extremely important that the workforce are adequately skilled to understand IoT and ML outputs. Through regular training programs and hands-on workshops geared to the engineers and decision-makers, it helps them understand the data which they can intelligently use for highly effective utility of the systems.

Collaboration with Experts: Incorporating IoT and ML would be challenging in terms of technical implementation, and a collaboration between geotechnical engineers, data scientists, and IT experts will address this challenge. It enables for system wide optimization, from sensor placement to model construction.

10.8 FUTURE DIRECTIONS AND ADVANCES

It seems likely that the tools for slope stability monitoring in mining of the future will be based on a blend of new devices connected through IoT combined with ML and DL to interpret data. 5G communication networks, for instance, offer great promise in terms of real-time data transmission in remote mining environments. 5G allows more efficient connection of IoT sensors to monitoring systems with high bandwidth and almost no latency to reduce the demand on processing data efficiently. Furthermore, Edge AI is gaining in popularity as well where data from IoT devices can be processed near the edge of a network, encouraging less use of cloud-centric systems and permitting on-site decision-making faster.

Similarly, another aspect here is drone-based monitoring systems with fusion of IoT sensors and even ML algorithms. This means that these drones are capable of capturing geospatial data in difficult or hazardous conditions, enabling slopes to be monitored for stability from the air within areas where conventional sensors may not be practical. Drone technology coupled with ML models provides more logical basis for slope stability and improves the accuracy in risk prediction by identifying the interaction of environmental and geological factors.

I would expect to see more advanced geotechnical sensors becoming part of the monitoring arsenal in the future, allowing us to interpret conditions well below the surface. For instance, sensors that can monitor soil properties and groundwater levels in real-time to even micro-seismic activities give mining operators a better understanding of slope stability controlling factors. Smart sensors, with self-diagnostic and self-calibration capabilities, require minimal maintenance leading to more reliable data over the long term.

The other significant innovation has involved implementing AI in intelligent systems that make autonomous decisions. These systems will also learn continually from the data they are collecting and make small incremental updates to their predictive models in real-time. For example, AI-based systems can more accurately predict slope failures by examining historical data to identify recurring indicators of unstable land. These models are expected not only to generate alerts but also to provide automated response recommendations, instructing operators how to deploy countermeasures like strengthening buildings or shutting down operations — in order to avoid accidents.

Another hot use case of blockchain technology is for securing and validating geotechnical data. Blockchain can eliminate various bad actors and ensure integrity of IoT sensors, which is critical to prevent tampering in the data captured from these sensors, which is best used for tracking efforts, compliance requirements to regulatory bodies, or operational audits.

The mining industry, in particular, is likely to have IoT and ML technology that become more entrenched facets of geotechnical engineering practices over the years. Ultimately, IoT networks paired with AI will develop such a fully automated real-time monitoring system that slope stability models would dynamically modulate through continuous updates and refinements. These systems will offer mining operators real-time esponses and answers to proactive actions, reducing the need for human involvement nd greatly lowering slope failure hazards.

It explained that as ML algorithms continue to become more advanced, the accuracy f our predictions and the risk assessment models will improve based on an extensive et elaborate understanding of the local and global environment in terms of operational actors. The use of multimodal data sources that combine geological, meteorological, nd operational data will certainly improve the predictive capabilities of these systems, esulting in more accurate forecasting on slope instability events.

Digital twin's 3D digital models of physical mining environments are expected o be developed and take slope stability monitoring a step further. The digital twins rought to life by IoT and AI will mimic in-the-field conditions as they happen and get ven sneakier with future slope movement. This technology will enable mining operaors to run a variety of scenarios and measures in an interactive simulation environment o that decisions can be more informed, while reducing the risks involved.

Sustainability will shift the focus to green IoT technologies that lessen power usage f sensor networks and ML systems. With environmentally conscious mining compaies, it is only those energy-efficient IoT solutions that will practically maintain coninuous monitoring without sharply increasing on energy consumption.

There will, finally, be shared data and learning between loads from multiple mine ites and geosystems over collaborative platforms that map shared intelligence. These latforms will enable access to enormous datasets that could be used by ML models for urther training and enhance slope stability predictions at different scales both across eographical and geological contexts.

10.9 CONCLUSION

Ensuring flat slope stability on undeveloped cast mines is still a critical issue with major implications for safety and operation efficacy, in the field of geotechnical engineering. By entering the ML coupled with IoT, we have solution that interprets connected world-generated data in unique ways. We can use IoT sensors and devices to collect land condition data in real-time, like no other we have seen before, ensuring that our slopes are monitored at all times. When IoT is of hybrid with ML, it empowers the creation of more advanced predictive models that can forecast slope failure prior to happening. By doing so, using this more sophisticated modelling, we can take the complex geotechnical data, which exist and overlay on it, the patterns, and the characteristics that could give us a signal of instability reach our site operators sooner than traditional methods – thereby being able to perform actions prior that were not yet possible based on current systems today. But applying this technology is hindered by several challenges on the way. Addressing technical challenges such as system integration issues, data quality, and data processing issues in a largely innovative manner followed by best practices. Lessons from the case studies demonstrate both the promise and limits of deploying those technologies in practice. Summary: The future of slope stability monitoring is bright as new technologies are being developed, and IoT and ML continu to evolve. New advances are on the horizon that will continue to transform our pra tices, creating opportunities to increase safety and efficiency of mining activities. Wi greater use and exploration into these technologies, the vision of better slope stabili solutions in cast mines could become a reality as we move forward to a safer future i geotechnical engineering.

REFERENCES

Chen, Z., Li, C., Hu, X., & Zhao, B. (2020). A hybrid machine learning model for slope stabili prediction: Applications in open-pit mines. *Natural Hazards*, *101*(3), 953–969.

Das, S. K., Basudhar, P. K., & Chakraborty, A. (2017). Slope stability assessment using artif cial neural networks (ANN) and geotechnical parameters. *International Journal of Minin Science and Technology*, *27*(1), 155–165.

Gholami, R., & Zarei, M. (2018). Smart IoT-based systems for real-time monitoring of slop failures in open-cast mines. *Sensors*, *19*(9), 2201.

Ghorbani, M. A., Zarei, M., & Gholami, R. (2020). Integration of IoT and machine learning i slope monitoring for enhancing safety in mining industries. *Journal of Applied Science an Technology*, *33*(4), 357–372.

Hassan, M. A., & Mahmoud, A. (2021). Machine learning models for slope stability predictior using IoT data. *International Journal of Mining Science and Technology*, *31*(1), 98–109.

Kumar, R., & Kaur, J. (2022). Innovations in IoT applications for predictive slope failure pre vention in mining. *Journal of Geotechnical and Geoenvironmental Engineering*, *148*(2 1033–1044.

Li, S., & Fang, H. (2021). Real-time data analytics with IoT-based machine learning for slop stability in mining operations. *Automation in Construction*, *124*, 103605.

Madani, N., Gholami, R., & Wetherelt, A. (2020). IoT-enabled machine learning techniques fc monitoring slope movements in mining. *Computers and Geotechnics*, *123*, 103618.

Sharma, M., Jha, M. K., & Das, R. K. (2021). Machine learning for predicting slope stabilit in mining activities. *Journal of the Southern African Institute of Mining and Metallurgy* *121*(1), 21–29.

Singh, A., & Sharma, K. (2021). Advanced IoT-based safety systems for slope stability i open-cast mining. *Mining, Metallurgy & Exploration*, *38*(2), 517–526.

Wang, Y., & Li, Z. (2019). Risk assessment of slope stability using machine learning algorithms *Geomatics, Natural Hazards, and Risk*, *10*(1), 2303–2315.

Zhang, K., Li, P., & Zhou, G. (2019). IoT-based remote monitoring system for slope stability anal ysis in open-pit mines. *IEEE Transactions on Industrial Informatics*, *16*(3), 2364–2374.

Zhang, L., & Wei, Q. (2021). IoT-integrated machine learning model for open-pit mine slop analysis. *International Journal of Mining, Reclamation and Environment*, *35*(2), 89–101.

Zheng, S., & Tang, H. (2020). Application of IoT and AI for real-time slope monitoring and analy sis in large mining operations. *Engineering Geology*, *271*(1), 105632.

11 Internet of Things-Based Approach Employing Examining for Probable Landslides

11.1 INTRODUCTION

Landslides are proven to be natural disasters and bring extensive damage to infrastructures along with loss of human life. These events are commonly cited to be initiated by a number of environmental and geological factors involving heavy rainfall, seismic activity, or erosion. When a landslide happens, the destruction left behind highlights the importance of monitoring systems that can promptly predict landslides and reduce their consequences. The rising risk of these types of disasters in landslide-prone areas as urbanization and infrastructure development carries on highlights the need for novel solutions.

11.1.1 Landslides and its Impacts

There is no predictive way of estimating landslides – they come unannounced and they are treacherous. When activated, these pose a threat to the owners and ultimately lead to property loss, infrastructure damage, or inevitably, some form of deaths. Landslides also

DOI: 10.1201/9781003537625-11

cause significant disruption to daily living life, closing roads, damaging homes, and costing communities economically, particularly in regions with unstable terrain. Moreover, landslides can transform landscapes forever even in a level of an eco-region destroying ecosystems and altering biodiversity. Due to the high expense associated with landslide recovery and reconstruction, both engineers and policymakers must prioritize solutions for prevention before landslides occur or for early detection of potential failure.

11.1.2 How IoT Is Transforming Geotechnical Monitoring

In the field of geotechnical engineering, one of the most important changes is how we work: Thanks to IoT technology, and new possibilities have opened up for carrying out these tasks, in a way that before it was literally impossible. The IoT devices collect readings from multiple sensors located in slipping zones; these data play an important role for landslide monitoring. These devices include biaxial tiltmeters, extensometers, in-place inclinometers, load cells, and piezometers and are used to give important information on soil moisture, slope stability, and ground movement. By analysing this information, early warning signs of potential landslides can be identified so as to take action before a disaster occurs (Figure 11.1).

11.1.3 Detection and Prediction in Landslide-Prone Areas

Prevention of landslides greatly depends on early detection and timely prediction. This helps authorities in leading evacuations, warding off people from the paths, and barricading roads on time, thus, saving human lives and cutting down damage. This can be accomplished by having IoT-based systems continually monitor geotechnical properties and provide real-time data to engineers and disaster management teams (Mohan et al., 2023). Accurate predictions allow timely decisions to be made to save lives and protect infrastructure; this is particularly important in areas that are affected by landslides regularly.

11.1.3.1 Soil stability and rock movements

The major role of landslide prediction is soil stability. The sliding of a slope occurs because the forces acting on it exceed its potential strength, especially during times of additional stress like rainfall or seismic activity (Ali et al., 2022). Engineers can determine the probability of a landslide by watching how rocks and soil move. IoT sensors placed strategically in the field provide live reports on ground movement, which can be used to trace movements of the ground over time, providing insights about the dynamics of flow (Li & Zhang, 2020). These data allow engineers to create more accurate models to determine when and where the landslide is most likely to happen.

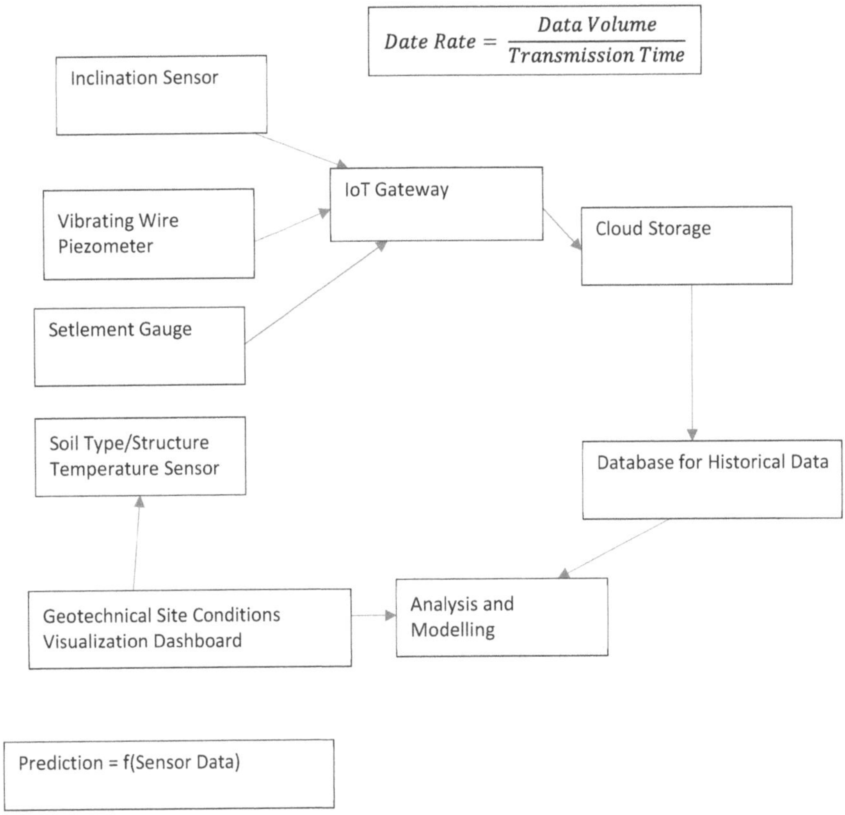

FIGURE 11.1 Internet of Things (IoT) transforming geotechnical monitoring.

11.1.3.2 Environmental factors: Rainfall, seismic activity, and erosion

Landslides are most often triggered by environmental factors such as rainfall, seismic activity, and even erosion. Extreme rainfall can saturate the soil, making it heavier and weaker on the slope, resulting in a debris flow. Even mild tremors can destabilize already precarious slopes, triggering landslides or avalanches, or turning hills into a pile of mud (Vishwakarma et al., 2024). Soil will also slowly erode, eventually weakening the structure and helping to create landslide-prone areas over the years. IoT systems detect these surrounding parameters in real-time and help in predicting the conditions for the possibility of a landslide.

11.1.3.3 Slope stability and groundwater impacts

Another very important attribute to take into account during landslide monitoring is slope stability. Nearly, all slopes contain water in some form or another, whether from rainfall runoff or continuous seeping into the surface of lakes and rivers. Groundwater can have a substantial impact on slope stability during times of high precipitation (Lynn et al., 2020). Excessive amounts of water accumulate, increasing the overall weight and causing a reduction in its capacity to withstand movement. IoT sensors that measure groundwater levels and soil moisture content are required to quantify water effects on slope behaviour. In this way, monitoring of water conditions in those areas enables engineers to better comprehend the influence of water on landslide risk and identify areas where preventive measures should be most effective.

In this case, sensors are 'in' the field measuring parameters related to soil movement, moisture changes, or ground vibration levels. There are wireless communication networks for the continuous transmission of data between these sensors and a central system (Kumar & Singh, 2021). With the help of these IoT devices, data are collected, and analysis is done to identify any patterns or possibilities that can result in an earth slide. Since the volume of land to cover is often huge and may not be easily accessible in some regions, IoT systems offer a compelling, scalable way to monitor vast lands.

11.1.3.4 Sensors for geotechnical engineering

The field of geotechnical monitoring uses diverse sensors, among others, to measure the physical environment. Tiltmeters, which track variations in the angle of slopes, allow researchers to understand soil displacement (Image: Amitrajeet Kumar). By monitoring these strains with strain gauges, a potential problem with the structure can be detected. Groundwater pressure has a huge influence on the stability of a slope, and piezometers are used to measure this. This helps reflect real-time data to engineers, supported by IoT networks; the stability of a concerned area can be monitored by these connected sensors, providing an integrated approach to landslide prevention through real-time interventions.

11.1.3.5 Sensor networks for remote sensing

A wireless sensor network (WSN) is the elementary building block of an IoT-based landslide detection system. These networks consist of numerous sensors that communicate wirelessly using an interconnected central controller. WSNs are also applied in locations difficult to access, where manual monitoring is a daunting task (Ali et al., 2022). In this way, engineers can monitor large areas of land in real-time without having to conduct physical inspections through networks of sensors. Efficiency at landslide monitoring rises manifold, helping in more precise predictions and avoiding loss of life and damage by providing early warnings.

11.1.3.6 Geotechnical parameters affecting landslides

Landslides are intricate events that develop through the interplay of several geotechnical factors. For efficient monitoring and prediction of landslides, it is important to understand the main factors generating instability in slopes. These factors include the stability of soil, movements of rocks, environmental forces such as rainfall and seismic loadings, and groundwater effects on slope stability (Vishwakarma et al., 2024). To reate the best estimate of where landslides are likely to occur, scientists need to understand these variables that contribute to landslide risk assessments; then early warning ystems can be built, saving lives and infrastructure.

11.2 SOIL STABILITY AND MOVEMENT OF ROCKS

The most important type of landslide risk is soil stability. Slides occur when soil is either too wet to bind itself or when a force such as more so for our domicile in National Highway (NH), being the weight of sodden roof snow pushes against it. There are many factors that affect the stability of the soil, and some of them include how it was formed, what type of compaction it underwent, and its moisture content. Rock layers under the soil are also important in slope behaviour. Rock movements which are natural, e.g., tectonic mobilizations or artificial like by construction operations, etc., can disturb the slope and reduce its structural strength. Using IoT sensors for monitoring soil and rock stability (e.g., tiltmeters and strain gauges) enables engineers to spot early indications of potential landslides.

11.2.1 Environmental Controls: Precipitation, Seismic, and Erosion

In many instances, landslides are trigged by environmental factors. One of its main causes is heavy rainfall, which saturates the soil and takes away its strength as it increases in weight. Too much water in the soil makes it easier for the ground to slip off. The shaking itself, even of low magnitude, is possibly able to unload weakened slopes by rapid horizontal redistribution of mass, reinforcing landslide activity in areas already subject to loss of strength for a variety of reasons. For example, erosion may weaken the supporting structure on a slope through slow progression and causes the slope failure. These types of environmental factors are monitored by real-time IoT systems, providing much needed data for early warnings.

11.2.2 Stability with Respect to Slope Failure and Groundwater Influence

The significance of groundwater in relation to the stabilization of slopes is also high. An increase in water within the soil will lead to a greater overall weight of the slope, and it will reduce the internal strength of soil, making it more susceptible to collapse. Groundwater pressures can cause slip planes, which are created when the soil pulls apart and slides downslope. Piezometers and IoT sensors also need to be installed order to monitor water content in soil which is essential for knowing if there are ar groundwater level changes that could compromise a slope. One key to is the interactic between groundwater and slope stability — if we know how much a given portion water content, underlies movement determines the potential for collapse inside lan slide area.

11.3 IoT TECHNOLOGY IN LANDSLIDE MONITORING

The application of Internet of Things (IoT) in geotechnical monitoring has change the prediction and management of landslides forever. From a technology integratio point of view, IoT assists engineers and decision-makers to determine potential lanc slides by providing an automated mechanism for real-time data collection, analysi and communication. Such proactive monitoring would prevent extensive damage t critical infrastructure and the loss of human lives. IoT systems have three key con ponents — sensors, communication networks, and data analytics — that work i unison to provide real-time field insights long after installation at even hard-to-reac locations.

IoT stands for Internet of Things — a network of physical devices, sensors, an systems that collect and exchange data over the Internet or specialized communicatio networks. An IoT system for landslide monitoring includes sensors inside the ground t measure geotechnical parameters, such as soil moisture, ground movement, and atmc spheric parameters. Targeted sensors also connect to remote centralized or cloud-base systems that enable remote real-time monitoring of high-risk locations. In landslid detection, IoT components generally consist of sensors for collecting data, wireless ne works for transmitting the data, and a data analytics platform to interpret and proces the collected information.

11.3.1 Sensors Used in Geotechnical Engineering

Many different sensors are used to detect key parameters in landslide monitoring, whic could represent an approaching sliding slope. Tiltmeters, for example — also employe by the Prentice Institute at Castle Rock and Cardiff University in Wales — are use to measure subtle movements in a slope's angle that might indicate ground movemen Strain gauges in the sidewalls measure deformations — stresses within the soil and roc layers, which are a precursor to failure. The groundwater pressure and water conten of the soil, an important factor to consider in relation to the probability of landslides are measured with piezometers. These systems incorporate sensors that have bee programmed to monitor constantly, alerting the system of any change in soil or environmental conditions as soon as it occurs (Chen & Zhang, 2019; Singh & Kumar, 2020).

11.3.2 Remote Monitoring with WSNs

Landslide monitoring by means of WSNs is crucial to observe landslides in remote areas, like those where manual observation and data collection are difficult because of rough terrain or hard geographical conditions. A WSN is composed of several interconnected sensors that wirelessly communicate with a central hub or data storage system. This is really important in places like landslide-prone regions, where these networks can send data without needing any physical maintenance to be done or a human present all the time. WSNs employ communication technologies like Zigbee, LoRa, or 5G to ensure that data transmitted from sensors are reliably sent, even in harsh environmental conditions, so unattended coverage of large areas over a long period can be guaranteed (Hakim et al., 2024).

11.3.3 Landslide Detection Systems Based on IoT

An IoT-based solution for landslide detection can further add to the advantage in monitoring and forecasting landslides in real-time. A patterned architecture is used to design the IoT-based landslide detection system with numerous sensors, communication networks, data analytics, and decision-support tools. The systems monitor environmental and geotechnical parameters continuously, making it possible to detect and deploy preventive measures early. The key to efficient mechanisms like these is the capability of a built-in system to deliver unbroken, precise, and instantaneous data in reducing risks in landslide-prone areas (Gupta & Sharma, 2022).

11.4 ARCHITECTURE FOR THE IoT-BASED LANDSLIDE DETECTION SYSTEM

Landslide type IoT-based architecture consisted of three layers as the perception layer (sensors and actuators), network layer (the communication protocols), and the application layer (data analysis and visualization). In the perception layer, sensors are arranged to acquire geotechnical and environmental information, which is then transmitted through the network layer via wireless communication technologies like Zigbee, LoRa or NB-IoT. At the topmost position in the stack (application layer), data gets processed and analysed (using edge computing or cloud-based platforms) to give real-time insights and visualizations. Architected well for data flow and communication as a layered approach, which is ideal for predictive analysis in Figure 11.2.

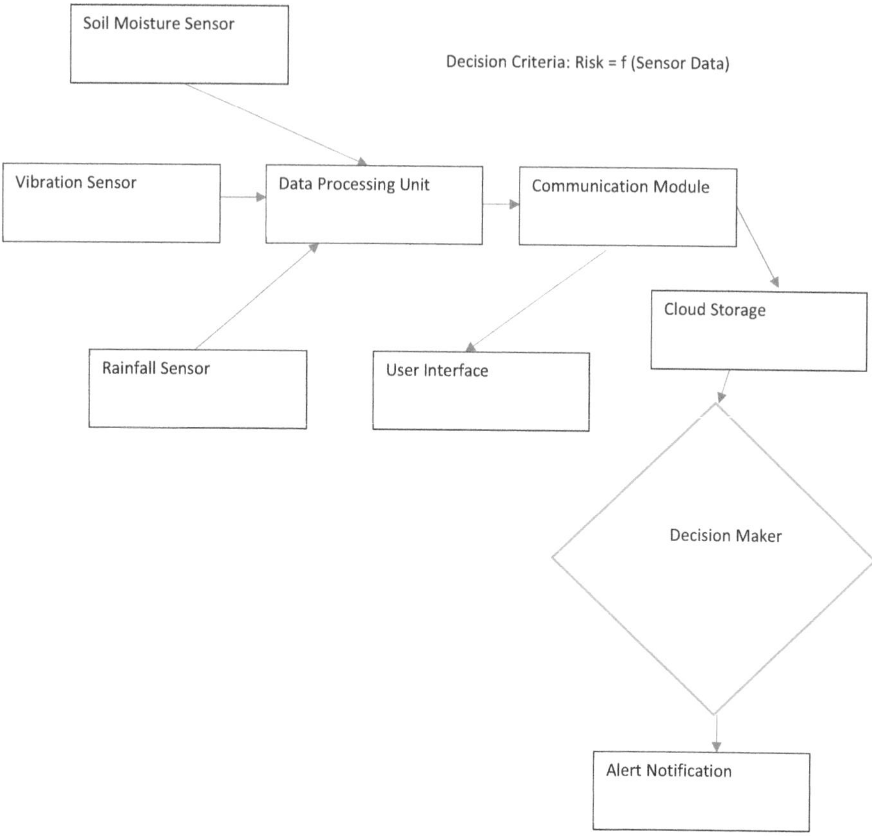

FIGURE 11.2 Architecture for the IoT-based landslide detection system.

11.4.1 Design of Sensor Node and Placement Strategy in Landslide Zones

An IoT-based landslide detection system largely depends on the design and deployment of sensor nodes, which play a significant part in ensuring accurate and timeliness of results. Sensor nodes are in general low-power devices that must be able to work reliably also under severe environmental conditions (e.g., high or low temperatures and humidity) and have a long service life. Many sensor nodes are deployed in the risk space of landslide areas, for instance, slope area and water seepage zone. They install a combination of sensors at as many nodes as possible, depending on topography, dispersed equally across various soil characteristics and the historic landslide patterns. It should be placed from appropriate places; this information is used to represent the condition of the area where landslide is happening and will provide correct detection in real-time.

11.5 DATA TRANSFER AND COMMUNICATION PROTOCOLS IN IoT SYSTEMS

To operate efficiently, an IoT-based landslide detection system requires data transmission that is both reliable and dependable. The success of the real-time monitor is directly dependent on connectivity choices for communication protocols and the enablement to maintain data flow in harsh environmental conditions. Since many systems are in remote locations or more obscure areas of operation, wireless communication technology is paramount for continuous data movement. Moreover, to ensure system reliability and operational efficiency, one must consider power consumption and the importance of edge computing (Patel & Desai, 2020).

11.5.1 Wireless Communication Technologies

In the case of landslide monitoring, each technology has its own advantages, and these are some common wireless communication technologies used in IoT systems:

LoRa (Long Range): It is best for long-range, low-power communication and, being so, is a good choice for remote landslide monitoring. It is able to transmit data over greater distances in the order of kilometres at minimal power consumption, yet with limited data throughput compared to other technologies (Zhao & Wang, 2019).

Zigbee: Another low-power, short-range technology best suited for sensor networks covering a wider area. However, its range is not as far reaching as LoRa, making it better for more local monitoring.

NB-IoT (Narrowband IoT): Created for long battery life and deep data connections, and NB-IoT can serve as a communication channel even in difficult areas like rocks and underground, which is why it is a good option to be coupled with landslide detection systems (Patel & Desai, 2020).

5G: When it comes to fast data transmission, low latency, and a higher number of successful connections, 5G is superior. However, its high-power requirements and limited availability in remote areas might make this an impractical solution for landslide monitoring systems (Elavarasi & Nandhini, 2021).

The specific technology used will depend on the location of the area being monitored, the type of data being measured, and the required power constraints.

11.5.2 Power Consumption and Connectivity

Identifying the terrain where landslide detection systems are employed is often rugged, isolated, and lacking real-time connectivity and a power source. Efficiently sending a data stream with low power consumption is essential for these setups. Conditions like these are perfect for low-power, wide-area networks (LPWANs) such as LoRa and NB-IoT, which use little energy and can provide considerable coverage. Sensors in remote areas are often powered by solar panels and energy-harvesting techniques to minimize maintenance visits.

Similarly, establishing reliable connectivity in adverse terrains (e.g., mountains) will require communication protocols capable of dealing with frequent signal loss. Using redundant communication pathways and mesh networks, where nodes transmit data through one another, ensures that some levels of information make their way back to the central system even when parts of the network have lost connectivity (Patel & Desai, 2020). Backup data storage at the sensor node level can prevent data loss in case of temporary connectivity failures.

11.5.3 Localized Data Processing Using Edge Computing

Landslide detection systems benefit greatly from edge computing, that is, the ability to process data at the level of sensor nodes or near the source of data. This provides a significant advantage in federated learning in IoT systems, where sharing large amounts of raw data for processing in centralized cloud servers could be time-consuming and bandwidth-heavy (Elavarasi & Nandhini, 2021). Edge devices perform real-time data analysis, and early alerts are triggered when a certain threshold is exceeded, leading to faster decision-making without the need for cloud communication.

Such localized processing of data reduces the load on central processors and allows operation in extremely remote or dynamic terrains, minimizing latency and power consumption as well. Edge computing allows time-sensitive alerts — critical for landslide monitoring, where quick response is vital — to still be issued if Internet connectivity with the cloud is lost.

11.6 REAL-TIME DATA ANALYTICS FOR LANDSLIDE PREDICTION

The universally vital element of real-time data analytics for effective landslide prediction is at its core. Through the real-time processing of detectable environmental and geotechnical data, IoT systems will help in recognizing patterns before any landslides, which can then lead to valuable and timely warnings. Machine learning (ML) and advanced data analytics were successfully integrated to enhance the accuracy and rapidity of forecasting landslides. Such technologies help policymakers to act in advance well before a disaster happens, thus saving lives and resources.

11.6.1 Data Analytics and ML for Landslide Prognosis

The data analytics and the ML algorithm are important factors for the feature of predictive capability in IoT-based landslide monitoring systems. It uses ML techniques to analyse data collected by sensors in areas identified as landslide risk zones and is capable of detecting minor changes in factors like soil moisture, land movement, and environmental conditions detected from these sensors in or around the landslides. ML models, such as decision trees, neural networks, and support vector machines (SVMs), are trained with the historical data to detect patterns that signal a potential of land sliding. They are able to learn through new data and therefore eventually become much better in predicting, thanks for its ability to adapt the forecast based on recent information.

By enabling the automatic analysis of massive quantities of data that would be far too voluminous or time-consuming for human analysts to handle, ML significantly improves forecasts for landslides. They are also dynamic: As environmental features change, these models can incorporate the new data and refine their predictions.

$$\mathcal{P}_{\text{landslide}}(t) = \sum_{i=1}^{N} \left(\omega_i \cdot \frac{\mathcal{D}_i(t)}{1 + e^{-\gamma \cdot \mathcal{F}_i(t)}} + \lambda \cdot \frac{\partial \mathcal{R}_i(t)}{\partial t} \right)$$

$$\mathcal{D}_i(t) = \int_0^T \left(\alpha_i \cdot \frac{\partial h_i(t)}{\partial t} + \beta_i \cdot \mathcal{V}_i(t) \cdot \frac{\partial^2 \mathcal{H}_i(t)}{\partial t^2} \right) dt$$

$$\mathcal{F}_i(t) = \frac{1}{1 + \log(1 + \mathcal{S}_i(t))} \cdot \int_0^T \mathcal{G}_i(t) \cdot \frac{\partial \mathcal{C}_i(t)}{\partial t} \, dt$$

$$\mathcal{R}_i(t) = \int_0^T \frac{\partial^2 h_i(t)}{\partial t^2} \cdot \frac{1}{1 + \log(1 + \mathcal{S}_i(t))} \, dt$$

$$\mathcal{S}_i(t) = \sum_{j=1}^{M} \left(\frac{\mathcal{A}_j(t) \cdot \frac{\partial \mathcal{L}_{ij}(t)}{\partial t}}{1 + e^{-\delta \cdot \mathcal{V}_i(t)}} \right)$$

$$\mathcal{G}_i(t) = \frac{\partial \mathcal{L}_{ij}(t)}{\partial t} + \kappa \cdot \frac{\partial^2 \mathcal{L}_{ij}(t)}{\partial t^2}$$

$$\mathcal{C}_i(t) = \frac{\mathcal{H}_i(t)}{\log(1 + \mathcal{V}_i(t))} \cdot \left(\sum_{j=1}^{M} \frac{\partial \mathcal{A}_j(t)}{\partial t} \right)$$

$$\mathcal{A}_j(t) = \frac{\partial \mathcal{L}_{ij}(t)}{\partial t} \cdot \left(\frac{1}{1 + e^{-\theta \cdot \mathcal{V}_i(t)}} \right)$$

$$\mathcal{V}_i(t) = \frac{\mathcal{H}_i(t)}{1 + \log\left(1 + \mathcal{L}_{ij}(t)\right)} + \sum_{j=1}^{M} \frac{\partial^2 \mathcal{A}_j(t)}{\partial t^2}$$

$$\frac{d\mathcal{P}_{\text{landslide}}(t)}{dt} = \sum_{i=1}^{N} \left(\omega_i \cdot \frac{d}{dt} \left(\frac{\mathcal{D}_i(t)}{1 + e^{-\gamma \cdot \mathcal{F}_i(t)}} \right) + \lambda \cdot \frac{d}{dt} \left(\frac{\partial \mathcal{R}_i(t)}{\partial t} \right) \right)$$

$$\mathcal{H}_i(t) = \int_0^T \frac{\partial \mathcal{L}_{ij}(t)}{\partial t} \cdot \frac{\partial^2 \mathcal{L}_{ij}(t)}{\partial t^2} \, dt$$

$$\frac{\partial \mathcal{L}_{ij}(t)}{\partial t} = \alpha \cdot \mathcal{L}_{ij}(t) + \mu \cdot \frac{\partial^2 \mathcal{L}_{ij}(t)}{\partial t^2}$$

Symbol Explanation:

- $\mathcal{P}_{\text{landslide}}(t)$ is the landslide probability prediction at time t.
- $\mathcal{D}_i(t)$ is the data analytics function for sensor i over time t.
- $\mathcal{F}_i(t)$ is the feature extraction function for sensor i.
- $\mathcal{R}_i(t)$ is the risk assessment function for sensor i.
- $\mathcal{S}_i(t)$ is the slope stability function for sensor i.
- $\mathcal{G}_i(t)$ is the gradient function related to landslide variables.
- $\mathcal{C}_i(t)$ is the critical load factor for landslide determination.
- $\mathcal{A}_j(t)$ is the sensor j acceleration factor.
- $\mathcal{L}_{ij}(t)$ is the landslide load factor between sensors i and j.
- $\mathcal{H}_i(t)$ is the hazard function for sensor i.
- $\mathcal{V}_i(t)$ is the vulnerability index for sensor i.
- N is the number of sensors.
- M is the number of measured variables.
- $\alpha, \beta, \gamma, \delta, \kappa, \lambda, \theta, \mu$ is the model parameters and regularization coefficients.

11.6.2 Environmental and Geotechnical Data: Pattern Analysis

Now when it comes to landslide prediction, we have to do the same things with two type of data: Environmental and geotechnical. Informative data like rainfall intensity, temper ature, and environmental changes are largely responsible for determining the likelihoo of a landslide. Data analytics help to reveal patterns, such as extended periods of heav rain or rapid temperature changes that might put a region at risk of landslides. Finally public-facing tools and apps can execute fine-grain messaging about the threat level i any given area. In the same way, when compared to rainfall accumulation rates or ground vibrations (from geological sensors), geotechnical information such as soil moisture content and groundwater flow velocities help to explain what is happening below the surface (and hence why it might become more likely for a landslide-producing event).

Time-series analysis and correlation studies are advanced data analytics techniques, which can help identify the early warning signs of instability seen in a landscape. For instance, rapid uptake of soil moisture associated with heavy rains may indicate saturation and loss of structural integrity of the soil. Likewise, slow ground and rock layers' minor movements measured by tiltmeters or strain gauges over time interval can give forewarning of potential slope failure. Landslide monitoring systems can issue early warnings with high precision if environmental triggers are related to geotechnical responses.

11.6.3 Operationalizing Early Warning Predictive Models

Early warning systems rely on predictive models. Predictive models are built for determining the risk of a landslide based on IoT sensor data and real-time ML to determine if it will slide, combining current with historical data. Using large dataset, the model gets trained to predict future events by recognizing these thresholds or patterns that lead to landslides. For instance, if levels of soil moisture exceed a critical threshold and ground movement speeds up, the system could send out an early warning allowing authorities to respond.

They also provide an alerting function: Pre-set levels of risk have been configured — for example, moisture content, rate of ground movement, or atmospheric conditions – whereby the predictive models deliver an automatic alert to identified interested parties. These thresholds are typically established based on data history and real-world testing, allowing for the system to be sensitive against risks whilst remaining robust against false alarms. In addition, some systems utilize ensemble learning methods — or in other words combine multiple prediction models to improve the depth of fewer lists and increase robustness.

11.7 CLOUD-BASED LANDSLIDE MONITORING PLATFORMS

Cloud-based platforms are used to centralize data produced by IoT-based landslide monitoring systems, enabling management, storage, and analysis of such significant amounts. Huddart said that these platforms allow for easy interfacing of IoT data with high level computational tools and access to geotechnical and environmental information at real-time from remote locations. To improve the accuracy of operational effectiveness for landslide forecasting systems, cloud-based operation and analysis based on real-time monitoring data are considered useful methods. Further, real-time dashboards and visualizations give decision-makers crystal clear actionable insights to make timely interventions (Table 11.1).

TABLE 11.1 Cloud-based landslide monitoring platforms

FIELD	*COMPONENT*	*DESCRIPTION*	*ADVANTAGES*	*DISADVANTAGES*	*USE CASES*
Data collection	Sensor networks (geophones, inclinometers, etc.)	Sensor networks deployed in landslide-prone areas to measure critical parameters like ground movement, soil moisture, vibrations, and rainfall	Provides real-time data on environmental and geological conditions, helping predict landslides early	High deployment costs, especially in remote and difficult-to-reach locations	Early warning systems in landslide-prone regions, real-time monitoring in mountainous, and rural areas
Communication network	Wireless communication protocols (LoRa, 4G/5G, etc.)	Facilitates data transmission from the deployed sensors to the cloud platform, using wireless protocols based on geographic needs and data volume	Enables real-time or near real-time data transmission even in remote regions	Unstable connectivity in remote areas and high energy consumption for continuous data transmission	Remote landslide monitoring in rural or isolated areas with minimal infrastructure
Cloud platform	Cloud infrastructure (AWS, Google Cloud, etc.)	Centralized cloud infrastructure that collects, processes, and stores data from sensors, allowing remote access and monitoring from anywhere	Scalable infrastructure capable of processing large amounts of data and providing high availability	Dependence on Internet connectivity; high operational costs if the cloud infrastructure is extensive	National and international monitoring centres, disaster risk management agencies, universities, and research labs

(*Continued*)

TABLE 11.1 (*Continued*) Cloud-based landslide monitoring platforms

FIELD	*COMPONENT*	*DESCRIPTION*	*ADVANTAGES*	*DISADVANTAGES*	*USE CASES*
Data processing and analysis	Artificial intelligence (AI)/machine learning (ML) algorithms for prediction	ML models and AI algorithms analyse historical and real-time data to predict the likelihood of landslides based on environmental trends	Can automatically learn from data and improve accuracy over time, reducing the need for manual intervention	Requires significant amounts of historical data for accurate predictions, which may not always be available	Predictive analytics in landslide-prone areas, alert systems for early disaster detection, and public safety monitoring
User interface and alert system	Dashboard and mobile applications	Visual interfaces and mobile apps that provide stakeholders with real-time updates, data visualizations, and automatic alerts for landslide risks	Accessible from anywhere with an Internet connection, providing instant alerts to multiple stakeholders	Alert fatigue if too many warnings are issued, and possible delays due to poor Internet connectivity	Governments, disaster response agencies, local communities, and emergency management services

11.7.1 Internet Data to Platform Internality

The basis of modern landslide monitoring systems is the integration of IoT data using cloud computing. They collect important readings, such as soil moisture, ground movement, rainfall, temperature, etc., which are essential in a landslide-prone area for prediction and monitoring purposes. This information is then sent to cloud servers to be processed and interpreted in real-time (Kumar & Singh, 2021). Data can be centralized in the cloud so that information from different sensors and locations can be accessed through one system.

IoT data volumes grow, but the architecture will remain performant as the volume increases due to scalable data storage and processing in cloud computing (Li & Zhang 2020). The cloud can host ML algorithms and predictive models for ongoing data analysis and the creation of early alerts. These integration elements ensure that the system maintains its efficiency and responsiveness whilst being ready to handle challenging and diverse data streams from different source types.

11.7.2 Cloud Storage for Long-Term Retention and Analysis Integration

Cloud-based platforms provide the ability for long-term analysis to be conducted by retaining large volumes of historical data. Landslides typically result from long-term changes in surrounding environmental or geotechnical conditions, so measuring subsidence rates over many years is crucial for understanding any changes (Thirugnanam et al., 2022). It maintains past data (e.g., previous landslides, sensor readings, and weather patterns) in cloud storage, so this historical context is always accessible for analysis.

Saving past data for the long term enables pattern recognition and more accurate predictors. ML algorithms can detect trends over long periods that are not often visible when examining only short-term datasets. Then, cold storage is a secure and dependable form of data backup that daily backs up important information against local system crashes or disasters. It also creates valuable collaborations between researchers, engineers, and decision-makers by allowing access to historical data from any point globally.

11.7.3 Real-Time Dashboards and Decision-Support Visualizations

A pivotal component in cloud-based landslide monitoring platforms is real-time dashboards and visualizations. The visual interpretation of information gathered from IoT sensors is quickly understood by decision-makers, providing helpful insights using these tools. Dashboards visualize real-time sensor data and predictive models alongside historical data in an easily interpretable way, empowering proactive decision-making and early warnings of increasing landslide risk.

Ground movement, soil moisture, and rain intensity are some key parameters that can be tracked with visualization tools comprising heatmaps, graphs, and trend lines. Operators can visualize the data in real-time and watch for thresholds being met, signaling a potential landslide, thus providing alerts to advance evacuation or remediation. Dashboards enable various teams to collaborate, allowing instant access to real-time data on multiple devices, which is useful for better coordination during critical matters.

Cloud-based dashboards also commonly include automated reporting capabilities, such that the insights delivered by predictive modelling and data analytics are often provided in the form of reports to support decision-making. These reports can be tailored for geography, time scales, or risk factors of interest — providing stakeholders with the necessary information to address landslide mitigation and response strategies.

11.8 IoT-BASED LANDSLIDE DETECTION SYSTEMS: CASE STUDIES

There are valuable lessons to be learned from the case studies of IoT-based landslide detection systems, which demonstrate how these technologies have been adopted around the world, and the real-life problems they aim to solve. This will help geotechnical engineers and decision-makers alike to understand how IoT has been successfully used to lower landslide risks, as well as to implement early warning systems. These case studies additionally show its services the lessons of previous deployments as guidance for future implementations.

IoT-based landslide monitoring systems have been successfully implemented at various sites in different parts of the globe with unique geotechnical problems. In the Himalayas of India, where landslides are very common and often bring livelihoods and infrastructure to their knees, IoT sensors track soil stability, rain support, and ground movement. Hence, real-time WSN and ML-based landslide prediction systems have evolved in present-day technical scenario.

Also in Japan, land is particularly susceptible to landslides due to a combination of regular heavy rains and frequent seismic disturbances on the crustal plates beneath. Encouragingly, similar IoT-led systems have been deployed here: Using digital cameras mounted on poles that are remotely activated by rainfall intensity. These are combined with arrays of tiltmeters, piezometers, and strain gauges to monitor ground movements and soil saturation in affected areas. Remote and bare environments can be realised due to the data collected by these sensors, being transmitted using NB-IoT and LoRa communication technologies, which have impact on real-time monitoring.

The system has been used in Latin America, specifically the Andes mountai to help protect local communities from landslides caused by a heavy combination rainfall and seismic activity. In the municipality, IoT-based monitoring systems are int grated with satellite communication in this region to address connectivity challenges high altitude terrains and stream data non-stop for advance warnings.

11.8.1 Success Stories

The hallmark of IoT-based systems for monitoring landslides lies in early warning ar loss-life prevention. For example, in Italy, a landscape IoT system on an area of lanc slide predicted the biggest that might occur and still swipe people off including nume ous lives. Edge computing was used for localized data processing, and ML models we then deployed to spot patterns in soil movement and local environmental conditions.

Massive IoT networks have already been established in, for example, Sichuan i China that is highly prone to post-earthquake landslides. The new system, incorpora ing both measurements taken in real-time from seismic sensors and geotechnical mon toring devices, was successful at predicting landslides induced by aftershocks. An earl warning system that so far is credited with saving lives by giving notice to authoritie before volcanoes erupted has been established.

Yet these implementations have also run into myriad problems. The supply of powe and communication connectivity are the major handicaps in remote areas. The effect c severe environmental changes on decommissioned systems has caused data leaks in th communication stream. Also, keeping sensors accurate to know that the data gathere are reliable can be deemed nearly impossible in hard-to-reach places.

11.8.2 Existing Deployments

The case studies on successful deployment of an IoT-based landslide detection systen First, you have to do is install the sensors; accuracy is everything. Deploy sensors based o the terrain and suspected landslide-prone zones, ensuring coverage of all possible geotech nical parameters like soil moisture content, groundwater levels, and ground movement.

Second, overcoming communication obstacles in rural areas necessitates carefu choice of technology — namely, LoRa and satellite IoT — to transmit data continuall Often, edge computing (with its companion cloud platforms) was indispensable in man aging bulk data and thus removing much network congestion to process everything fo real-time alerts without any time lag.

Finally, it is also a major point of consideration because scaling the system an keeping cloud cost at minimum will determine how mass tree performs. This wa whilst large systems can be monitored extensively for impending trouble, Manohar sai smaller, low-cost deployments too can be used to offer an effective early warning sys tem. This balance between accuracy, scalability, and cost is crucial for devising sustain able and impactful systems.

11.9 CHALLENGES IN IoT-BASED LANDSLIDE MONITORING

There are different issues to be solved in order to guarantee the reliability and efficiency of IoT-based systems for landslide monitoring. This is especially true in remote or hostile terrains where natural conditions can make sensor operation difficult, data transmission challenging, and system scalability an issue. Insight on how to work through these challenges is needed in order to advance the science of predicting and preventing landslides.

11.9.1 Power Supply and Energy Management

Strong, long-lasting power supply is one of the largest challenges to address in deploying IoT sensors to monitor landslides in remote places. In much of the landslide-prone regions, rugged and remote terrains and a semi-productive economy result in inadequate infrastructure, posing great challenges to supply power to sensors. Wired power solutions are often not feasible, and the latest trend is to rely on batteries, solar panels, or energy-harvesting alternatives.

Nevertheless, solar-powered systems have their own challenges, especially those designed for areas that receive a lot of rain or have weeks with lots of cloud, which can disrupt the normal collection method, so then one needs some form of storage. In addition, frequent battery replacements and/or recharges in harsh weather conditions cause additional overhead for maintenance efforts. Smart energy management strategies including low-power sensor design and efficient communication protocol are essential to extend the lifespan of each individual sensing node and keep them running without interruption.

11.9.2 Data Accuracy, Reliability, and Sensor Calibration

To provide early warnings, accurate predictions, accuracy, and reliability of IoT data in monitoring landslides are necessary. But sensor calibration can be extremely difficult, especially in areas affected by weather or conditions that could change over time (like soil erosion or seismic activity). Fine-scale soil creep and rising groundwater levels, monitored by tiltmeters, strain gauges, and piezometers must be measured extremely accurately in order to trigger the flooding warning system.

It is possible to achieve good data at the outset, but over time, it can get pretty challenging due to sensor drift, environmental wear, and tear as well as the accumulation of dirt around sensors. Thus, one needs to periodically recalibrate it; however, the access of remote sensor locations is not always feasible. To begin to address these problems, automated calibration methods both in the sensor architectures and self-correcting algorithms along with ML models that can compensate for sensor drifts are being researched.

11.9.3 Remotes and Difficult Terrains with Communication Problems

Real-time landslide detection and warning requires efficient communication between sensors and central monitoring systems. But many landslides happen in remote, mountainous regions with weak or no wireless communication signals. These signals can also be blocked by terrain obstructions such as cliffs, dense vegetation, or rock formations, and weather condition like heavy rain or snow can disrupt wireless transmissions.

To solve these issues, IoT-based landslide monitoring systems usually employ a mix of low bandwidth and long-range wireless technologies such as LoRa and NB-IoT that provide far reach data transmission with constrained energy consumption. To reach remote areas, satellite communication or mesh networking is sometimes used for reliable connectivity. Nonetheless, at the edge level, providing constant bus data flow is still a task as environment may be adverse, and network latency can reduce real-time monitoring purpose due to the delay of seconds on each packet-fetch operation.

11.10 SECURITY AND PRIVACY IN IoT FOR GEOTECHNICAL APPLICATIONS

Integration of IoT with geotechnical applications like landslide monitoring pose several hurdles related to security and privacy considerations. As such data are integral to forecasting and managing such catastrophes, it is of paramount importance to maintain the integrity, confidentiality, and availability of that information. The stability and effectiveness of IoT-based geotechnical solutions rely heavily on vigilant adherence to data integrity protection and secure transmission, as well as safeguarding against cybersecurity threats, ensuring reliable real-time monitoring, accurate data analysis, and seamless integration with advanced engineering systems.

11.10.1 Cyber Security Threats to IoT System

The IoT system, specifically that is strategically placed in remote locations such as landslide-prone areas, is threatened by a variety of cyberattacks. These threats involve unauthorized access, data manipulation, denial-of-service (DoS) attacks, and malware propagation. For example, cyber criminals could target vulnerabilities in IoT devices or communication protocols to access high-impact systems without permission and interfere with the monitoring process or counterfeiting data.

Since it is so important that the data from landslide monitors remain accurate, any breach of security could lead to false alarms that may result in significant risks to human safety. Consequently, cybersecurity features need to be hardcoded into the design of IoT solutions, and that they should include at least robust authentication mechanisms, encryption, or sufficient consideration for later deployments of security patches. Intrusion detection systems and routine security checks help to identify and neutralize threats on the network.

11.10.2 Data Assurance and Access Control for Landslide Monitoring

The importance of data integrity is crucial in landslide monitoring system as it is prerequisite for the timely warning and proper decision-making. The integrity of data is to protect it from unauthorized changes and maintain its accuracy from the time it is collected until it goes in transmission or stored.

Access control mechanisms are implemented to limit access to the sensitive data and system components. This includes best practices for setting up roles and permissions to users, using multi-factor authentication and limiting changes to systems and data from authorized individuals only. Using standard audit logs and access reviews you shall keep track on who gets the data and why so you significantly reduce inside threat. Encrypting the data is another crucial step toward ensuring data integrity. To avoid unauthorized access and manipulation, you should encrypt both data at rest (stored data) and data in transit (data being transmitted). Encryption algorithms need to be refreshed at a regular cadence as new threats appear between updates.

11.10.3 Securing and Reliable Data Transfer

To the success of the above IoT-based landslide monitoring systems, reliable and safe data communication is equally imperative. Data must be transmitted from remote sensors to central systems securely, and it must not intercept, lost or tampered with.

Secure communication protocols over-the-air encryption (TLS or SSL) VPNs for securing connections between sensors and central systems. Moreover, message integrity checks and LOFEC can improve the reliability of transmitted data, making sure that only full and correct packets are available for processing.

These strategies in combination with redundant communication paths and buffering techniques can be used to alleviate challenges associated with high-latency or intermittently connected environments, such as remote/on-premises deployment scenarios in rugged terrains where connectivity is the main concern. Performing these initial tasks at the edge decreases the volume of data that need to be transmitted and reduces risks associated with data interception as well, two tangible benefits to both security and system efficiency.

11.11 AFFORDABILITY AND SCALABILITY OF IoT LANDSLIDE EARLY DETECTION SYSTEMS

Introducing IoT links to landslide detection systems must find the balance between low cost and scaling up. Although the potential benefits of such systems in terms of early detection and risk mitigation are significant, effective financial and operational management are key to success. This section specifically looks at difficulties in budgeting large-scale deployments, preserving system scalability and cost effectiveness, and achieving the optimal balance between accuracy and cost efficiency.

11.11.1 Large-Scale Guidelines: Budgeting and Resource Allocation

If one thing does occur, that another and very serious issue is the budgetary compulsion for huge-scale enactment of IoT-based landslide detection systems. The associated costs (e.g., sensors, communication infrastructure, data processing platforms, and maintenance) can be significant. Sensors are one piece of the equation. Many additional factors such as sensor installation, site preparation, and ongoing operational costs will play a role.

A well-designed budget must be provided for these costs, both in the initial set-up and maintenance. Spending on the highest leverage areas might be worthwhile as is pursuing sensor technologies that are cost-effective. You could also get support from government grants and funding from disaster management programs. It is important to allocate resources intelligently in order to monitor the right areas effectively, without over spending. Where should we allocate our resources — are strategic planning and cost-benefit analysis required? Haugen drew an analogy that is the system of using more advanced heat sensors in high-risk zones as compared to cheaper but less effective solutions across some lower-risk areas.

11.11.2 System Scalability for Larger Area

Especially when implemented in areas with variable levels of risk, scalability is a factor for IoT landslide detection systems that must be considered. Scalable systems can grow with the size of a network or increase capacity for coverage without excessive reconfiguration or costly out-of-pocket. This means designing the system with additional sensors and more data processing nodes when required.

One way you can provide scalability is through modular system design. It is also built to be expandable incrementally through standardized sensor units and communication modules. The advantage of a cloud-based platform is that it can easily scale with storage and the ability to process larger quantities of data, without significant changes in infrastructure.

Another challenge we face is the scalability problem, which includes solving network capacity and data management problems. To address this, some approaches could be implemented as efficient data aggregation techniques with distributed computing can help in handling high volumes of data consuming from large-scale sensor networks. This also helps to scale when it integrates with the existing infrastructure and interoperate with other monitoring systems.

11.11.3 Trade-offs between Cost and Accuracy in IoT Systems

Establishing how to balance cost versus accuracy is a fundamental part of designing IoT-based landslide detection systems. The better the sensors and technologies, the higher the accuracy of course but these might also result in more expensive tools. On the contrary, cost-effective solutions can slightly miss the target but provide more extensive concentration.

It can be cost-effective by delivering the most value where it is more crucial to being right. As an example, the budget can be optimized by deploying high precision sensors in high-risk areas that could have a large number of people getting affected if a landslide happened, whilst less expensive sensors could be used in low-risk areas. Hybrids: Mix GNSS and non-GNSS data source enhance global system accuracy more economically.

These approaches need to be augmented by using data fusion techniques that will integrate data obtained from various sensors and sources to further improve accuracy. These data can in turn be analysed by ML algorithms, which provides more accurate predictive abilities and lower false alarms. System evaluation and performance assessment should be conducted regularly to ensure that the system's balance between cost and accuracy remains relevant as the system progresses.

11.12 FUTURE TRENDS IN LANDSLIDE MONITORING USING IOT

As technology progresses, this area of IoT-supported monitoring for landslides is expected to see a lot of innovation. Technology has been revolutionising the way geotechnical monitoring is done — and continues to do so with such innovations as artificial intelligence (AI), blockchain, and advanced sensors. There will be integration with satellite IoT and automated systems as well, to greatly enhance landslide detection and prediction facilities.

11.12.1 Emerging Technologies: AI, Blockchain, and Advanced Sensors

Advanced data processing and predictive analytics enabled by AI are the cornerston that will revolutionize the way landslide monitoring is conducted in future. AI alg rithms are able to handle large amounts of data from multiple disparate data source and they have the capability to spot trends and outliers that any traditional methods m easily overlook. For example, ML models can examine past landslide data combine with current sensor readings to produce better forecasting and early warning system As time goes on more data to apply will be available, helping those models will impro all the time in order to deliver better predictions.

Blockchain technology has great potential to address these challenges and contribu towards improved security and reliability of landslide monitoring related data. These da collected from sensors are records of data transactions, so blockchain helps in makin the transaction immutable and traceable. This is paramount to maintain confidence the consistency and resolution of landslide predictions. Furthermore, on blockchain pla forms, smart contracts can trigger automated responses to threshold events; for example, can issue alerts or invoke mitigation measures when thresholds have been reached.

The progress in landslide monitoring is being powered by advanced sensors well. Advances in sensor technology, like fine resolution tiltmeters, fibre optic sensor and smaller GPS units, have made measurement of ground movement and water conte more accurate. The durability of these sensors is on the rise, and they have also prove to be effective in harsh environment conditions, to increase the level of service provide by monitoring systems.

11.12.2 Satellite IoT Spatio-Temporal Geotechnical Monitoring

Satellite IoT contributes very largely in geotechnical monitoring by delivering the wid landscape of landslide susceptible areas. Advantages of the satellite technology are th enormous coverage area, very frequent imaging, and topographic details. Using rada and optical sensors mounted on satellites, the satellite can identify ground deformatio and changes in land cover that indicate increased likelihood of a landslide.

Incorporation of satellite data in conventional IoT-based monitoring systems improve the overall landslide detection efficiency. Remote sensing data, for example, can be use in conjunction with readings from ground-based sensors to more fully understand th environmental conditions that lead to landslides. Additionally, it provides real-time moni toring and early warning system, which allows alerts and action in no time.

These satellites can also come in handy for disaster response and recovery as the can deliver valuable feedback on the areas struck by disasters. This information ca be invaluable to coordinate rescue activities, estimate the damages, and plan for futur preventive actions.

11.12.3 The Future of Autonomous Systems in Landslide Detection

The deployment of autonomous systems is expected to transform the detection and monitoring of landslides, providing state-of-the-art capabilities for data acquisition, analysis, and response. Drones, for instance, are already being put to use conducting aerial surveys of landslide hotspots — they carry high-definition cameras, capturing images and data from locations that are otherwise hard to access. A wide range of sensors such as cameras, LiDAR, or environmental sensors can be attached to these drones for an exhaustive data about the ground conditions.

Automated ground vehicles and robots are also being designed to remotely monitor and upkeep sensor networks in difficult terrains along with the drones. Orchestration: These systems can operate in isolation from humans, doing things like calibrating sensors, collecting data, and maintaining the system with no human intervention.

These systems can be helpful for collecting data by integrating with autonomous systems so the analysed data will help the early detection of any potential landslides and provide a real-time response to that. Said systems can be configured to automatically change their surveillance parameters, append extra sensors, or trigger emergency protocols according to some specific pre-agreed conditions.

11.13 CONCLUSION

More recently, the uptake of IoT technology in geotechnical monitoring has led to great leaps forward in prediction and prevention of landslides. Since landslides can be so deadly and destructive — hazardous to human life, infrastructure, and ecosystems — the capacity for early detection and prediction is essential to save lives, reduce economic loss from damage to infrastructure and structures, as well as mitigate other environmental impact. IoT-enabled real-time data captured from soil sensors measuring stability, rock gauges registering movement, environmental condition including information on rainfall and seismicity allow us to have a better understanding of the behaviour of slopes. Such high-risk areas can be monitored using IoT sensors such as tiltmeters, piezometers, and a wireless network which in agreement with geotechnical engineers can create the best deal. These systems gather large volumes of data, and when run through advanced analytics and ML models, they are capable of predicting landslides in advance, so that policy measures can be taken accordingly. Cloud-based platforms make it even better by having very secure and scalable ways of storing enormous amounts of data for long periods, making things more efficient with cool looking dashboards, visualizations, etc. IoT-based landslide detection also has challenges like power supply management, data accuracy, and communication in remote terrain but its advantages are far more than these difficulties. AI, blockchain, and satellite IoT shall make the landslide monitoring more system accurate at scale and affordable levels. In the extended future as IoT matures, it will gradually take over the reins in preventing landslides with its omnipotent grasp on all things and guarantee an even safer haven to those endangered by landslides.

REFERENCES

Ali, O., Ishak, M. K., Bhatti, M. K. L., Khan, I., & Kim, K. I. (2022). A comprehensive review of Internet of Things: Technology stack, middleware's, and fog/edge computing interface. *Sensors*, *22*(3), 995. https://doi.org/10.3390/s22030995.

Chen, J., & Zhang, H. (2019). A novel IoT-based landslide monitoring system with edge computing. *Sensors*, *19*(10), 2345. https://doi.org/10.3390/s19102345.

Elavarasi, K., & Nandhini, S. (2021). Landslide monitoring and tracking using IoT sensors. *Journal of Physics: Conference Series, 1717*, no. 1, 012060. DOI:10.1088/1742-6596/1717/1/012060.

Gupta, R., & Sharma, P. (2022). Real-time landslide prediction using IoT and big data analytics. *International Journal of Disaster Risk Reduction*, *60*, 102345. https://doi.org/10.1016/j.ijdrr.2021.102345.

Hakim, W. M. A., Ramli, A. F., Basarudin, H., Abu, M. A., & Ahmad, I. (2020). WSN and IoT based landslide monitoring system. *Test Engineering and Management, 83*, 10926–10932. DOI:10.7753/IJSEA1304.1006.

Kumar, A., & Singh, M. (2021). IoT-based landslide monitoring and early warning system: A review. *Journal of Environmental Management*, *287*, 112307. https://doi.org/10.1016/j.jenvman.2021.112307.

Li, H., & Zhang, Y. (2020). IoT-based landslide monitoring system with real-time data analysis. *Sensors*, *20*(15), 4321. https://doi.org/10.3390/s20154321.

Lynn, T., Endo, P. T., Ribeiro, A. M. N. C., Barbosa, G. B. N., & Rosati, P. (2020). The Internet of Things: Definitions, key concepts, and reference architectures. In Theo Lynn, John G Mooney, Brian Lee, Patricia Takako Endo (Ed.), *The Cloud-to-Thing Continuum* (pp. 1–22). Springer. https://doi.org/10.1007/978-3-030-41110-7.

Mohan, A., Dwivedi, R., & Kumar, B. (2023). IoT for landslides: Applications, technologies and challenges. In *Paradigms of Smart and Intelligent Communication, 5G and Beyond* (pp. 245–259). Singapore: Springer Nature Singapore. DOI: 10.1007/978-981-99-0109-8_13.

Patel, A., & Desai, M. (2020). IoT-based landslide detection and alert system using machine learning. *Procedia Computer Science*, *167*, 1234–1245. https://doi.org/10.1016/j.procs.2020.03.123.

Thirugnanam, H., Uhlemann, S., Reghunadh, R., Ramesh, M. V., & Rangan, V. P. (2022). Review of landslide monitoring techniques with IoT integration opportunities. *IEEE Journal of Selected Topics in Applied Earth Observations and Remote Sensing, 15*, 5317–5338. DOI:10.1109/JSTARS.2022.3183684.

Vishwakarma, A. K., Chaurasia, S., Kumar, K., Singh, Y. N., & Chaurasia, R. (2024). Internet of things technology, research, and challenges: a survey. *Multimedia Tools and Applications*, 1–36. DOI:10.1007/s11042-024-19278-6.

12 Patterns and Developments in Internet of Things (IoT) Investigation

Future Efforts in Geotechnical Engineering

12.1 INTRODUCTION

The construction industry has entered a new era of unprecedented technological change and innovation with the introduction of the Internet of Things (IoT). IoT offers a new way of doing things in construction engineering: it changes the game by providing real-time data mining, predictive analytics, and automation of tasks for better process efficiency and project performance (Katiyar & Kumar, 2021). How IoT trends and patterns in construction amplify the future of geotechnical engineering is significant. This chapter traverses deep into the pattern and evolution of IoT in construction, a sub-domain wherein others have been relatively meek to tread, including our one today as well. The merging of IoT and construction is not just a trend on the horizon, but a reality that must happen for there to be an increase in efficiencies and improvements in safety, security, and sustainability (Waqar et al., 2024).

DOI: 10.1201/9781003537625-12

Bear in mind, the previous chapter was an introduction to the subject of IoT as applied specifically within construction engineering. Via the connected devices and sensors, IoT in construction engineering creates an environment where a project can be monitored round-the-clock by deploying these devices for the measurement of many different parameters (Statsenko et al., 2023). From equipment usage and environmental conditions to building structure, IoT systems ensure data moves uninterrupted between machines, workers, and managers. With this connectivity, better decisions can be made and site management—as well as productivity—improved. This means better monitoring of construction sites, hence less human error and more resource utilization that will automatically result in faster project delivery time, enhancing safety standards.

12.1.1 Significance of IoT in Geotechnical Engineering

Geotechnical engineering is critical in keeping both construction projects safe and stable. IoT data for soil health includes real-time data reporting on soil conditions, moisture levels, and structural loads (Bianconi et al., 2019). These insights have allowed engineers to analyse where risks currently exist, forecast potential failures or bottlenecks, and take action before they escalate into real-world problems. From subsurface monitoring during tunnelling to the evaluation of environmental factors on foundations, IoT provides geotechnical analysis with an extra boost, improving results and ensuring safer constructions for all stakeholders (Al-Saeed et al., 2020).

12.1.2 Research Trends in IoT-Related Technologies

The IoT landscape is an ever-shifting sea, so there are probably a few trends and technologies mired in whispers as we sail forth to the drink, for a future construction engineer. IoT applications increasingly have involved wireless sensor networks (WSNs), cloud computing, and big data analytics. Moreover, the advent of edge computing has driven real-time processing at or near the construction site, thereby reducing data latencies and allowing quicker decisions to be made. Additionally, 5G networks are expected to amplify IoT application possibilities by offering the large bandwidth and low-latency wireless communication necessary for sophisticated and data-heavy construction processes (Alaloul et al., 2020).

The most important use of IoT in construction engineering among them is structural health monitoring. Engineers can now outfit buildings, bridges, and just about anything else with IoT-compatible sensors to monitor the structure in real-time. These systems monitor for characteristics such as vibrations, strain, temperature, and displacement to gain important insights into the health of structures, while also being able to do so in real-time. This approach helps to proactively monitor and identify issues early on, allowing for the prevention of expensive long repairs or, even worse, catastrophic failures. This also facilitates adherence to safety norms and prolongs the life of structures.

12.1.2.1 Construction automation with IoT

Automation in construction has been made possible by IoT, which allows repetitive and hazardous tasks to be implemented by networked machines. While drones, robotic bricklayers, and autonomous vehicles for construction can all be deployed individually today due to advances in technology, they are beginning to become part of a broader package that includes an IoT framework. These systems can work 24/7, ensuring accuracy, thereby minimizing downtimes and improving productivity. IoT automates typical manual tasks, speeding up the construction process while decreasing accidents on a site and boosting worker safety (Kanan et al., 2018).

12.1.2.2 Drones/remote sensing in on-site inspections

It became funny as a drone with IoT sensors attached during the construction site inspections. They literally give administrators a view from above the project to show real-time information about site conditions, structural advancement, and safety issues in Figure 12.1. When applied to construction sites, remote sensing technologies can enable in-depth topographical mapping, accurate site surveys, and even material volume measurements. By shortening the time for inspections and thus decreasing the

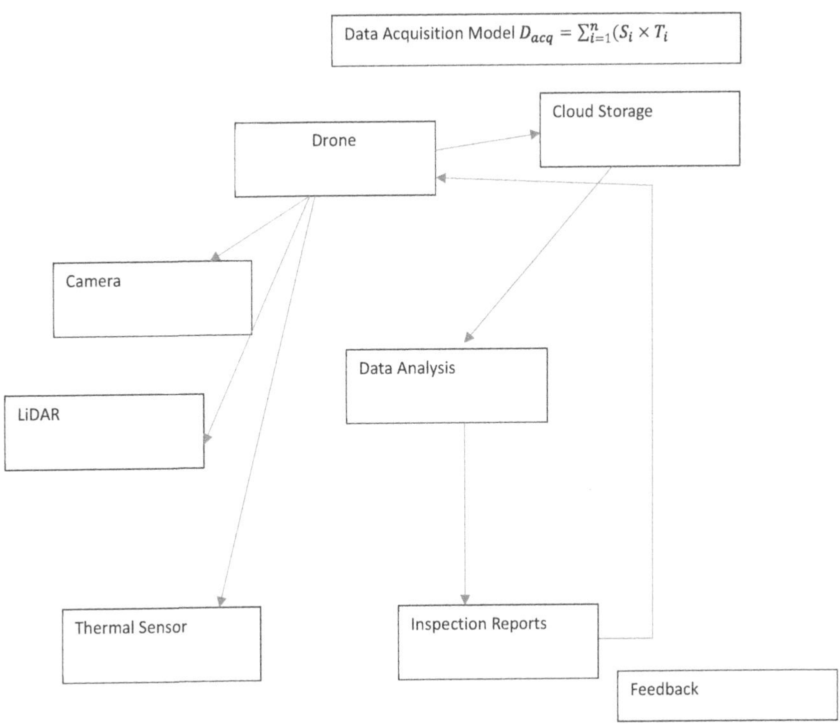

FIGURE 12.1 Drones on-site inspections.

costs of labour, associated with larger accuracy and coverage of their collected dat drones overall enhance decision-making through every stage of construction.

The advancement of technology has drastically evolved geotechnical enginee ing, with the incorporation of IoT into the field. Real-time, in-situ continuous monitc ing of subsurface conditions is made available by exploiting WSNs (Bilal et al., 2016 Cloud-based platforms store and process vast data produced by these sensors and edg computing enables rapid analysis of data as it happens (Karmakar & Delhi, 2021). E enabling engineers to find and address problems faster in the project lifecycle, the technologies will help make construction work safer and more cost-effective.

12.2 RECENT APPLICATIONS OF IoT IN CIVIL ENGINEERING

The use cases of IoT technology today hint at its future transformative power in co struction engineering. These new innovations are transforming construction site ma agement, data collection and analysis, and risk mitigation strategies (Darko et al., 2020 Technologies using IoT bring real-time data insights and automation, thereby enhancin construction safety, efficiency, and sustainability. Here are the top areas where IoT applied in construction engineering (Anding et al., 2013).

12.2.1 IoT-Enabled Structural Health Monitoring Systems

Structural health monitoring systems enabled with IoT are important when it comes to th safety and longevity of infrastructure. These are devices that are installed on importar structural areas, such as bridges, tunnels, and high-rise buildings, to constantly collec data regarding parameters such as strain, temperature, vibrations, and stress (Karmaka & Delhi, 2021). For this purpose, real-time data insights are used which provide informa tion when any degradation or potential failures occur so that maintenance can be cor ducted proactively before the point issue turns into huge critical alarms. It increases th longevity of structures and eliminates or reduces downtimes and overheads for the owner such that construction becomes more environmentally safe (Dale et al., 2019).

12.2.2 Automation in Construction with IoT

Automation: IoT is essential in the creation of automated processes on the construc tion site. The optimization of operations for equipment that was previously completel manually driven can now work as a large-scale IoT-based machine layer, from roboti bricklayers to autonomous excavators and cranes (Kannus & Ilvonen, 2018). They hav the ability to work with high precision (meaning fewer errors), thus boosting productivity and reducing labour costs. IoT automation also improves security by reducing the need for human presence in dangerous areas, therefore greatly lowering the possibilities of accidents occurring due to human error.

12.2.3 Development of Environmental Monitoring and Sensing System

IoT in construction engineering finds another vital application, such as environmental monitoring, especially in the case of geotechnical projects. For example, smart sensors can follow environmental parameters that include temperature, humidity, soil moisture, and air quality. These sensors allow project managers and engineers to monitor how ambient conditions might affect construction activities and materials (Bilal et al., 2016). A waterproofing system company also has to monitor the groundwater level during excavation to keep soil stability secure from sudden problems, such as water infiltration or landslides. Smart sensors also help construction projects maintain compliance with environmental regulations and serve to reduce the project's overall ecological footprint (Kang, 2021).

12.2.4 The Unmanned Skies: Drones and Remote Sensing in Site Inspections

Combining drones with other remote sensing technologies, such as IoT, has changed the way we inspect and survey our sites. Drones with IoT sensors provide high-resolution images and videos of construction sites, providing engineers with a 24/7 view of the project status. Technologies in remote sensing, such as light detection and ranging (LIDAR), will allow the construction of highly accurate 3-dimensional models of this job site which help in planning, risk estimation, and progress monitoring. This eliminates the manual system of inspection, which saves time and money by providing a new level of accuracy and safety. Remote monitoring of sites, especially in dangerous or remote locations, provides an added layer of security and ensures issues are addressed quickly.

12.3 TECHNOLOGICAL ADVANCEMENTS IN IoT FOR GEOTECHNICAL

The application of IoT in geotechnical engineering has greatly advanced the industry by providing ultra-accurate data acquisition, real-time interpretation, and predictive insights enhancing the outcomes of construction. In addition to improving safety and efficiency, these advancements have facilitated more sustainable and intelligent construction. There are some most important technological innovations in IoT that are reforming the geotechnical applications (Table 12.1).

TABLE 12.1 Technological advancements in Internet of Things (IoT) for geotechnical applications

FIELD	*DESCRIPTION*	*ADVANTAGES*	*DISADVANTAGES*	*USE CASES*
Smart sensors for soil monitoring	Advanced sensors equipped with IoT capabilities for real-time monitoring of soil parameters (moisture, temperature, etc.)	Enables continuous data collection, improving decision-making for soil management	Initial installation costs can be high; sensor calibration and maintenance may be required	Agricultural monitoring, construction site assessments, environmental monitoring, and land reclamation
Wireless communication protocols	Utilization of advanced wireless communication protocols (e.g., LoRaWAN, NB-IoT) to transmit data from sensors to cloud systems	Facilitates long-range communication with low power consumption, ideal for remote areas	Limited bandwidth can affect data transmission speeds; may require multiple gateways for coverage	Remote soil monitoring stations, landslide detection systems, and underground infrastructure monitoring
Data analytics and machine learning	Leveraging data analytics and ML algorithms to analyse geotechnical data for predictive maintenance and risk assessment	Enables predictive insights and improved accuracy in decision-making; reduces human error	Requires large datasets for training; complex algorithms may need specialized knowledge to implement	Predicting soil behaviour, assessing stability of slopes, and optimizing construction practices based on soil data
Geographic information systems (GIS)	Integration of IoT data with GIS technology for spatial analysis of geotechnical data and visualization	Provides valuable insights through spatial context, improving project planning and management	Data integration can be complex; requires specialized software and training for effective use	Urban planning, flood risk assessment, and geological hazard mapping
Blockchain for data integrity	Using blockchain technology to ensure data integrity and security in IoT systems, maintaining a tamper-proof record of data	Enhances trustworthiness of data, especially in regulated industries; ensures data provenance	Implementation can be complex; high resource requirements for maintaining the blockchain infrastructure	Securing geotechnical data in construction projects, environmental monitoring, and compliance tracking

12.3.1 Wireless Sensor Networks in Geotechnical Data Collection

For geotechnical data collection, WSNs have emerged as an indispensable instrument. These are a set of sensor networks which can be deployed all over the construction site and they keep on analysing the basis of soil humidity, temperature, pressure, and also the instant seismic activity. WSNs facilitate immediate data transfer, in this way providing engineers with remote access to information for analysis. An example of this is that because the user is able to pull in more complete high-resolution data, this reduces the potential for human error to occur due to manual data collection. WSNs also improve the detection and response to those hazards, such as soil movement or foundation conditions, at the initial stage of construction (Arashpour et al., 2017).

12.3.2 Real-Time Data Analysis Cloud-Based Platforms

Introduction to cloud computing and its amazing benefits of storing and analysing data has been a game changer for IoT-based geotechnical applications. The geographic information systems (GIS) solution deployed in the cloud offers a premium management of all that data which an IoT network can collect on site. These tools help analysis on a real-time data scope so that determination can be accessed from any location and by the project workers at any level. The cloud also enables team collaboration by making sure all key stakeholders are able to access the updated data and make well-informed decisions in a timely fashion. In addition, the data generated by continuous monitoring can be vast so cloud storage would provide you with scalability to store all that data as well, thus, enabling the use of advanced analytics tools to derive more value from this information.

12.3.3 Construction Sites Driven by IoT Data Edge Computing

The advantages of cloud computing are numerous, but there is no question that the increasing volume of data that IoT systems generate raises demand for edge computing in geotechnical engineering. Edge computing manages data on-site locally at the construction site, leading to less latency and bandwidth because it saves transmitting any type of data to cloud central servers. This helps in real-time analysis of crucial data that enables faster decision-making at the point of business. For instance, edge computing can provide instantaneous access to sensor data about the stability of the soil or performance of equipment so engineers can make on-site tweaks immediately, without waiting for a cloud-generated report.

12.4 INCORPORATING AI AND MACHINE LEARNING IN IoT SYSTEMS

Artificial intelligence (AI) and machine learning (ML) are playing a key role in the IoT world by streaming it with predictive analytics for decision-making and geotechnical engineering is no different. By comparing them with similar data and monitoring threshold limits, the AI algorithms are more likely to anticipate variations or failures which can streamline construction activity. Predictive maintenance and risk mitigation ML models can be used to predict early signs of equipment failure, soil instability, or structural stress helping in bringing about proactive or necessary engineering changes that can help before a condition develops. At the same time, AI-enabled IoT systems are able to take over basic tasks from human engineers and allow them more time to concentrate on their further experience of solving challenges ensuring that construction projects void any bottlenecks.

12.4.1 Case Studies: IoT for Modern Construction Projects

Deployment of IoT technologies on the modern construction sites has significantly transformed the way infrastructure is constructed, operated, and monitored today. This paper also presents some case studies from around the world that show what IoT can do to reduce risks in geotechnical and construction engineering, focusing on safety and productivity improvements. These live examples give us the cues on how IoT is solving the difficult-to-crack challenges in construction, making it easier for further advancement.

12.4.2 Smart Infrastructure Projects: A Worldwide Overview

The other side was that smart infrastructure projects were beginning to spring up around the world, where IoT drove both the construction process and future-proofing of infrastructure. One of the famous examples is building smart bridges and tunnels in countries like China, the USA, and the UK. These with IoT sensor-embedded infrastructure are monitoring parameters like vibrations, temperature fluctuations, and material fatigue regularly. The Hammersmith Bridge in the UK includes an IoT-enabled structural health monitoring system to quickly alert maintenance crews of signs of decay that need immediate attention. Across the globe, smart infrastructure initiatives are protecting against massive crashes and developing predictive maintenance models to extend the life of essential infrastructures.

12.4.3 IoT in Geotechnical Engineering for Risk Mitigation

Indeed, IoT systems enable these projects to yield major advantages tooled for geotechnical engineering studies—such as variability in soil condition changes by means of landslides or shifts carried due to earthquakes. Meanwhile, IoT so-called WSNs have already begun to be deployed in various landslide-prone areas of Japan (which muddy up things a bit, heh) to keep tabs on things like soil movement, moisture levels, and seismic activity. These data streams let engineers predict when landslides will occur and prepare for or possibly even avoid the worst risks before they happen. In a similar example from Hong Kong, the IoT is implemented for monitoring subsurface changes during excavations, preventing ground settlement, and maintaining the structural stability of existing neighbouring buildings. By integrating and scaling such IoT applications to geotechnical engineering, risks are reduced for construction projects and public safety.

12.4.4 IoT-Powered Worker Safety and Productivity

They are also making major contributions in optimizing construction site worker safety and productivity, In the construction industry nowadays, IoT technologies can likewise contribute a wealth of benefits. Smart helmets and vests as well: Wearable IoT devices like smart helmets and vests are being employed to track the health condition of the workers, which includes monitoring their heart rate, temperature, fatigue levels, etc. In a remarkable case study that we will discuss here, IoT wearable devices are attached to workers like other wearables and all their vitals remain monitored centrally on the cloud via an online platform; as soon as any vitals fall off the safe ranges, automated alerts get activated at respective passes causing immediate actions from corresponding supervisors to make certain safety of their people. IoT-enabled devices can also be used to monitor the location of workers at site, which in turn restricts access to certain restricted areas and minimizes accidents. The blending of these two skills health planning and the other location have decreased the number of accidents at the workplace a lot as well as built an improvement in overall performance.

12.5 IoT IMPLEMENTATION IN CONSTRUCTION ENGINEERING

The use of IoT technologies in construction engineering offers many advantages, yet it also poses some challenges that must be addressed to ensure their effective enforcement. These challenges play out across technical, operational, and security barriers—all of which present hurdles that must be surmounted before the industry can effectively exploit the full potential of IoT.

$$\text{CObjective:} C(t) = \sum_{i=1}^{N} \left(\alpha_i \cdot \frac{P_i(t)}{D_i(t)} + \beta_i \cdot \frac{E_i(t)}{T_i(t)} + \gamma_i \cdot R_i(t) \right)$$

$$P_i(t) = \int_0^T \left(\frac{\partial M_i(t)}{\partial t} + \lambda \cdot \frac{\partial^2 M_i(t)}{\partial t^2} \right) dt$$

$$D_i(t) = \frac{\int_0^T F_i(t) \cdot \left(1 + \log\left(1 + C_i(t)\right)\right) dt}{\int_0^T \left(1 + \log\left(1 + Q_i(t)\right)\right) dt}$$

$$E_i(t) = \frac{1}{\sum_{j=1}^{M} \frac{K_{ij}}{U_{ij}(t)}}$$

$$T_i(t) = \int_0^T \left(R_i(t) \cdot \frac{\partial R_i(t)}{\partial t} \right) dt$$

$$R_i(t) = \frac{1}{\sum_{k=1}^{L} \left(\frac{W_{ik}}{V_{ik}(t)} \right)}$$

$$\frac{dC(t)}{dt} = \sum_{i=1}^{N} \left(\alpha_i \cdot \frac{dP_i(t)}{dt} + \beta_i \cdot \frac{dE_i(t)}{dt} + \gamma_i \cdot \frac{dR_i(t)}{dt} \right)$$

$$\frac{dP_i(t)}{dt} = \frac{\partial}{\partial t} \left(\int_0^T \left(\frac{\partial M_i(t)}{\partial t} + \lambda \cdot \frac{\partial^2 M_i(t)}{\partial t^2} \right) dt \right)$$

$$\frac{dD_i(t)}{dt} = \frac{d}{dt} \left(\frac{\int_0^T F_i(t) \cdot \left(1 + \log\left(1 + C_i(t)\right)\right) dt}{\int_0^T \left(1 + \log\left(1 + Q_i(t)\right)\right) dt} \right)$$

$$\frac{dE_i(t)}{dt} = \frac{d}{dt} \left(\frac{1}{\sum_{j=1}^{M} \frac{K_{ij}}{U_{ij}(t)}} \right)$$

$$\frac{dT_i(t)}{dt} = \frac{d}{dt} \left(\int_0^T \left(R_i(t) \cdot \frac{\partial R_i(t)}{\partial t} \right) dt \right)$$

$$\frac{dR_i(t)}{dt} = \frac{d}{dt}\left(\frac{1}{\sum_{k=1}^{L}\left(\frac{W_{ik}}{V_{ik}(t)}\right)}\right)$$

$$\frac{\partial C(t)}{\partial t} = \sum_{i=1}^{N}\left(\alpha_i \cdot \frac{\partial P_i(t)}{\partial t} + \beta_i \cdot \frac{\partial E_i(t)}{\partial t} + \gamma_i \cdot \frac{\partial R_i(t)}{\partial t}\right)$$

Symbols Explanation:

- $C(t)$ is the construction efficiency at time t.
- $P_i(t)$ is the performance metric of IoT sensor i at time t.
- $D_i(t)$ is the downtime metric for sensor i at time t.
- $E_i(t)$ is the energy consumption of sensor i at time t.
- $T_i(t)$ is the time-based operational threshold for sensor i.
- $R_i(t)$ is the reliability metric for sensor i at time t.
- $M_i(t)$ is the measured data from sensor i at time t.
- $F_i(t)$ is the function of external factors affecting sensor i at time t.
- $C_i(t)$ is the connectivity measure for sensor i at time t.
- $Q_i(t)$ is the quality metric for sensor i at time t.
- K_{ij} is the cost parameter for IoT application j of sensor i.
- $U_{ij}(t)$ is the utilization factor for application j of sensor i.
- W_{ik} is the weighting factor for reliability of sensor i in environment k.
- $V_{ik}(t)$ is the value of performance metric of sensor i in environment k.
- N is the number of IoT sensors used in the construction project.
- M, L is the number of applications and environmental contexts respectively.
- $\alpha_i, \beta_i, \gamma_i$ is the weighting factors representing the influence of each metric.

12.5.1 Data Security and Privacy Issues in Section

Security and data privacy are some of the biggest problems with respect to IoT implementations along with dealing huge amount of generated data. Construction is increasingly becoming a target: construction sites house volumes of sensitive data including worker locations, health metrics, and structural data—all information that would be detrimental in the event of inappropriate access to the system. These devices are typically interconnected and have wireless communication features which cause them to be vulnerable to hacking or unauthorized access. In IoT-enabled construction environments, providing protection to personal and project-related data goes hand in hand with end-to-end encryption, regular security updates, and adherence to data privacy regulations.

12.5.2 Integration with Existing Infrastructure

However, whether equipping existing construction infrastructure with IoT systems or integrating into new builds from the ground up, it is a little more challenging than it should be, especially in older buildings and sites that were never intended to be hooked up to so much modern tech. A largely systemic issue across construction—many legacy systems simply can't work with IoT devices or platforms, necessitating huge retrofitting or operational expenditures. Moreover, construction organizations will also need to support integration with new IoT systems and solutions including pre-existing technologies like BIM or conventional sensors. Successful integration will require significant planning and fresh investment in new hardware and software.

12.5.3 Big Data Management and Analysis in IoT Systems

IoT generate huge amounts of intrinsic data, from time to time in real-time that is difficult for storage and analysis. Structural sensors, environmental monitors, and worker wearables are just some of the sources producing data from construction projects, yielding complex data streams that necessitate cutting-edge management solutions. This data is difficult to make into actionable insights because construction companies may lack the in-house expertise and tools needed for data analytics. Cloud-based platforms, edge computing, and AI-powered analytics can make a difference in the way big data is managed and processed within IoT products—but that does not mean it will be easy or inexpensive to implement these solutions.

12.6 IoT AND SUSTAINABILITY IN CONSTRUCTION ENGINEERING

The construction sector is answering the call for global sustainability and has begun to embrace the IoT as a technology that could help get better results. This will decrease the environmental spoor of construction companies, conserve resources, and enable constructive valuable output thereby achieving a circular economy. The way in which IoT is integrated into construction workflows contributes towards the use of sustainable practices to improve efficiency and develop creative solutions.

12.6.1 IoT for Reducing Environmental Impact

Perhaps the biggest advantage of the construction IoT is its capacity to reduce environmental impact through more disciplined resource management and monitoring in real-time. IoT-enabled systems help track energy spent, material used, and waste generated throughout construction. This allows companies to monitor equipment and processes, identify inefficiencies, and find ways to reduce fuel or minimize material waste. Two examples are IoT sensors that can be situated in construction equipment, monitoring emissions to ensure environmental regulations are met and reduced pollution and greenhouse gas output. With construction companies everywhere seeking to shrink their environmental footprint, the capacity to track and optimize resources is only getting more important.

Energy efficiency continues to move to the front of mind for developers, and IoT fixes a number of the problems energy management workers on construction sites face today. IoT smart energy systems allow lighting, heating, ventilation and air conditioning (HVAC) and other energy-intensive systems to be controlled offsite on construction sites. These real-time data-driven advanced systems reduce energy loss and operational expenses. The use of IoT sensors is not limited to the overall consumption but also it can be used to keep a tab on the machinery triggers making them require a potential maintenance before any wear and tear kicks in, thereby saving precious energy loss due to equipment damages. Furthermore, IoT implementations can make it possible to include renewable energy power sources like solar panels in the projects, thus improving energy efficiency and sustainability.

To put it simply, the circular economy is about optimizing material use through reuse, recycling, and reduction in order to ensure sustainability, and this is where IoT comes into play when it comes to construction. IoT technologies can follow the lifecycle of building materials making their own acquisition to disposal, preventing excess and waste. Whether or not concrete is a major contributor, IoT-empowered platforms can track the status of construction materials instantly and help to reveal reuse or recycle opportunities (e.g., for steel). For instance, in demolition projects, the IoT can track deconstruction allowing materials to be freed up and used in new projects. The consequence is a lesser demand for raw materials, and therefore waste deposited in the landfill will be reduced, contributing to those circular economy targets that matter as benchmarks in the construction queue.

The benefit of IoT in sustainability goes beyond just resource management to improving the efficiency and eco-friendliness of construction projects. Construction firms can boost their margins and help bring the greener future they claim to work for with IoT solutions down in their processes.

12.7 FUTURE DIRECTIONS TRENDS AND DEVELOPMENTS IN IoT FOR CONSTRUCTION ENGINEERING

Construction engineering is due to see a lot of transformative changes with the evolution of IoT technologies. IoT will enable improvements in predictive maintenance, digital twins, autonomous machinery, and secure data exchanges as the sector adopts digital innovations for safer, more sustainable, and efficient construction.

12.7.1 IoT-Driven Predictive Maintenance in Construction Equipment

IoT is bringing a revolution in the industry by implementing predictive maintenance using IoT on construction equipment management changes things. For example, the IoT sensors embedded in the machinery can monitor measures such as temperature, vibration, or running conditions to predict potential failures before they happen. With this data-driven process, construction companies know when their machines are most likely to need maintenance—eliminating the bad surprises of unplanned breakdowns and costly downtime. By pairing predictive analytics with IoT systems, construction teams can improve the longevity of equipment and generate long-term savings in maintenance or repair—all while streamlining project timelines. Predictive Maintenance: As IoT adoption increases, predictive maintenance will become the norm in construction.

12.7.2 Digital Twins in Construction and Geotechnical Engineering

Eager to replicate the success they experienced in industrial that would be difficult to recreate in real life but have an accurate, instant replica of it on a phone or tablet—and armed with design drawings, site data such as surveys and soil samples; project managers like Chapman started using digital twins. It is the data streams that feed into these digital models which IoT provides—using sensors in construction sites and equipment to continuously provide real-time data. Through digital twins, engineers can perform various what-if scenarios and takings, monitor the health of different pieces in their field, and create data-driven decisions throughout the life cycle of any given project. Digital twins in the domain of geotechnical engineering allows for a live monitoring of soil and foundation, helping to reduce risks and increase the accuracy of design. The increasing adoption of digital twins presents the potential to transform how construction projects are planned, performed, and maintained.

12.7.3 Construction Machinery Transformed into Autonomous by IoT

Autonomous machinery driven by independent technology solutions is the new way forward in construction. IoT sensors, algorithms, and ML can automate the operation of excavators, bulldozers, or even cranes remarkably independently. The machines can perform certain tasks with precision, save a lot of labour, and remove workers from dangerous environments that may be a part of any such process. IoT will have all the equipment on a construction site speaking as one in unison, thus converting any existing team to highly coordinated operatives where machines and crew can work seamlessly together. The technology has the potential to greatly increase the pace, cost-efficiency, and safety of construction projects as autonomous capabilities improve.

12.7.4 IoT and Blockchain for Protection Information Alternate in Construction

Construction: The IoT devices in construction will produce massive data sets and securing the integrity of this data largely becomes a concern. Combining blockchain technology with IoT provides a way forward for protecting the data transmitted on construction sites. Decentralized structures protect data recorded through IoT by securely storing and sharing information via blockchain so that no one sees or tampers with the data other than who has authority over it. The need to link critical path activities in a logical, sequential order is critically important as part of a network architecture, particularly for large-scale construction projects entailing many stakeholders (contractors, engineers, suppliers). By offering an immutable historical record of transactions, blockchain together with IoT can create greater transparency, traceability, and accountability in construction activities—a key driver for building trustful relationships and promoting collaboration among project stakeholders.

12.8 SMART CITIES AND IoT INTEGRATION FOR URBAN INFRASTRUCTURE DEVELOPMENT

Smart cities are changing the way our urban spaces are built and managed. The rise of the IoT in urban infrastructure is essential to building more connected and efficient, as well as sustainable cities. IoT also helps cities to solve infrastructure challenges, make operations greener, and generally improve quality of life by real-time monitoring and automation.

12.8.1 IoT for Smart City Developments

A key to smart city applications in general, IoT is a network of objects connected via the internet that can collect, analyse, and exchange data related to urban processes and functions. IoT can be used in cities to improve traffic management, control energy consumption, manage garbage collection, and maintain public safety. For instance, systems based on IoT can dim street lights according to pedestrian traffic, optimize traffic lights for better flow of traffic, and deliver live updates related to public transport. IoT optimizes urban services and improves the quality of life for citizens by enabling smarter and more responsive cities.

12.8.2 IoT for Urban Infrastructure Surveillance

IoT can keep checking the conditions of roads, bridges, tunnels, and water systems running under cities. These structures are fitted with sensors that monitor various parameters such as strain, temperature, and vibrations in real-time so that the condition of these structures can be understood throughout their lifespan. This also allows cities identify early signs of wear and tear, preventing from expensive repairs, while ensuri the safety of their public. IoT sensors can check the health of the earth and will mon tor the failure of infrastructure like when a darn due to be fall, it works in geotechn cal engineering so on. By being able to monitor the health of key infrastructure real-time, cities can better predict when maintenance is required, extend the life of a asset, and order around resources more efficiently.

12.8.3 Practical Engineering Application: Smart Cities and Geotechnical Solutions

IoT-based solutions are part and parcel of infrastructure development exercises th many cities the world over have already embarked upon. Some examples are smart cit projects in Singapore, Copenhagen, or Barcelona implementing IoT technologies fo intelligent traffic systems, water distribution, and efficient public buildings. IoT ser sors for IoT applications in geotechnical engineering are used to monitor landslide soil moisture, and underground water levels of the areas if there is any threat of natura disasters or long-term intervention. Moreover, IoT-based systems are also used to dete mine the sturdiness of earthquake-prone buildings. In this instance, the implication are more immediate as we can see real-world deployment of solid solutions that no only make smart urban development—including city waste management operations— possible but also play a role in furthering safety and sustainable infrastructure.

By establishing IoT with the urban infrastructure, smart cities will be able to detec problems before they occur and build a more sustainable living environment whil making sure it has hundreds of millions of people. The wedding of IoT and geotechni cal solutions means a future in which we can be better and safer and keep expandin the infrastructure of our cities.

Looking ahead, the question remains on how we could address this issue so thes IoT technologies contribute to the transformation of construction industry, by first con sidering necessary regulations and also ethical issues following along. The implementa tion of IoT systems in construction faces special difficulties due to an unbalanced degre of standards, data privacy, and worker rights; thus, they pose a concerted challenge tha demands careful management so as to derive maximum gain and minimize risk.

12.8.4 Addressing Worker Privacy in IoT Landscapes

In IoT-enabled construction sites, where devices are used to monitor many facets o worker activities and performance, much worry is over worker privacy. Are IoT tech nologies going to improve safety and productivity or are they tools of surveillance tha will allow companies to survey workers like never before? Privacy safeguards shoulc be put in place to ensure that only employees' work-related data is being collected, tha the data is securely stored and access thereto controlled, and the workers are apprised of why and how data would be monitored. There is a trade-off between using IoT for operational excellence and maintaining the privacy of workers. Conversations with employees and addressing their concerns are important steps in maintaining a respectful, transparent workplace while ensuring that privacy standards are met.

12.9 IoT—THE FUTURE OF CONSTRUCTION ENGINEERING

Construction engineering looks set to change in the future with advancements in IoT technology happening so fast. IoT is constantly evolving and has transformative potential to rewrite whole lanes of industry history books, bringing with it new efficiencies and breakthroughs as well as innovations in safety and sustainability. To make it through, this future and incorporate IoT innovations gained due prominence in construction implementation, collaboration among academia, industry, and governments is going to be a key component.

12.9.1 Anticipated Advancements in IoT Technologies

Construction engineering by its nature is on the cusp of fundamental transformation and IoT technologies could represent a major leap forward. Among the most awaited progress is the creation of more advanced sensors and devices with higher accuracy and performance. The sensors of the future may someday allow us to see minute-by-minute what a building's state is in all aspects, from structural health in earthquakes and wind-flow studies to dust concentration due to demolition spawning tropical algae. Further, adopting IoT systems integrated with advanced analytics, AI, and ML will empower predictive capabilities that predict errors even before they occur, thereby enhancing project management and bringing down downtimes. Added innovations such as blockchain for secure data sharing and advanced energy management systems will bolster the effectiveness intact of IoT, allowing for smarter and harder construction processes.

12.9.2 Collaboration among Academia, Industry, and Governments

For IoT to become a real-world ally of construction engineering, complex interlinkages between academia, industry, and governments need to be an ongoing effort. This demonstration of joint engagement is critical as academia continues to lead research and development in new IoT applications and technologies, even while they take application advances from the lab into the field. The governments are highly responsible for setting standards and regulations and sanctioning research projects. Those collaborative ventures could result in solid frameworks supporting IoT adoption, reducing regulatory bottlenecks, and planting roots for ethical framework deterrents. By bringing these stakeholders together at the program level, we can accelerate innovation and implementation, paving the way for new and innovative solutions to address complex construction challenges.

12.9.3 Pioneering the Future of Construction Engineering with IoT

IoT technologies will have an increasingly large impact on the future of construction engineering as they inevitably improve further. This will allow us to better align project outputs and operational efficiencies custom fit our requirements enabled by collected and analysed large volumes of data which collectively shall enable informed decision-making. This real-time data will help ensure the workers' safety and will help prevent any accidents or mishaps. Automation, prediction maintenance, etc., are some of the many operations that IoT-based solutions will provide to ensure the cost gets kept minimal. The penetration of IoT combined with other developing innovations like digital twins and self-autonomous machinery will take this to the further stage where it transforms the whole industry into smart construction environments which is more versatile and reactive. Adopting these innovations will not just change the way we build but also enable us to meet our future infrastructure needs with sustainability and resilience.

12.10 CONCLUSION

The implementation of IoT technology in construction engineering allows geotechnical challenges to be taken head-on in a far more advanced way. Structural health monitoring systems for improved safety and durability, the replacement of old-fashioned site inspections with drones and smart sensors—IoT is refashioning age-old customs. This amalgamation of IoT with the most advanced technologies like AI, ML, and edge computing makes it even more significant as it allows for faster data collection, real-time analysis, and even predictive analytics. Mobile sensor networks and wireless technology, as well as new concepts for geotechnical data management with cloud-based systems, are challenging the way things were done in how geotechnical risk can be managed more sustainably and efficiently. The increasing amount of smart infrastructure projects (worldwide) shows that the IoT has a place to overcome challenges like the limited interpretation of results and productivity. While these developments are a great leap forward, they bring with their own challenges data security, scalability, and interoperability are essential issues to be addressed if IoT is to realize its full promise within the construction engineering sector. Environmental sustainability, too is among the key pillars, that helps IoT to minimize adverse environmental impacts of such construction projects by increasing energy efficiency or achieving circular economies. Moving forward, developments in areas such as predictive maintenance, digital twins, and blockchain provide the next platform for innovation that promises to afford more power over construction processes and improved efficiency. As IoT becomes mission-critical in developing smart cities and infrastructure monitoring, the future of construction engineering is "connected" besides "smart" and "resilient." This would revolutionize geotechnical engineering, giving us a major breakthrough in the field if IoT can be made possible with the collaboration of academia, industry, and governments so that the ceiling is lifted to whatever extent.

REFERENCES

Alaloul, W. S., Liew, M. S., Zawawi, N. A. W. A., & Kennedy, I. B. (2020). Industrial revolution 4.0 in the construction industry: Challenges and opportunities for stakeholders. *AIN Shams Engineering Journal*, *11*(1), 225–230. https://doi.org/10.1016/j.asej.2019.08.010.

Al-Saeed, Y., Edwards, D. J., & Scaysbrook, S. (2020). Automating construction manufacturing procedures using BIM digital objects (BDOs). *Construction Innovation*, *20*(3), 345–377. https://doi.org/10.1108/CI-12-2019-0141.

Anding, K., Garten, D., & Linß, E. (2013). Application of intelligent image processing in the construction material industry. *ACTA IMEKO*, *2*(1), 61–73. https://doi.org/10.21014/acta_imeko.v2i1.100.

Arashpour, M., Bai, Y., Aranda-mena, G., Bab-Hadiashar, A., Hosseini, R., & Kalutara, P. (2017). Optimizing decisions in advanced manufacturing of prefabricated products: Theorizing supply chain configurations in off-site construction. *Automation in Construction*, *84*, 146–153. https://doi.org/10.1016/j.autcon.2017.08.032.

Bianconi, F., Filippucci, M., & Buffi, A. (2019). Automated design and modelling for mass-customized housing: A web-based design space catalogue for timber structures. *Automation in Construction*, *103*, 13–25.

Bilal, M., Oyedele, L. O., Qadir, J., Munir, K., Ajayi, S. O., Akinade, O. O., & Pasha, M. (2016). Big data in the construction industry: A review of present status, opportunities, and future trends. *Advanced Engineering Informatics*, *30*(3), 500–521.

Dale, P., Sporne, I., Knight, J., Sheaves, M., Eslami-Andergoli, L., & Dwyer, P. (2019). A conceptual model to improve links between science, policy and practice in coastal management. *Marine Policy*, *103*, 42–49.

Darko, A., Chan, A. P., Adabre, M. A., Edwards, D. J., Hosseini, M. R., & Ameyaw, E. E. (2020). Artificial intelligence in the AEC industry: Scient metric analysis and visualization of research activities. *Automation in Construction*, *112*, 103081.

Kanan, R., Elhassan, O., & Bensalem, R. (2018). An IoT-based autonomous system for workers' safety in construction sites with real-time alarming, monitoring, and positioning strategies. *Automation in Construction*, *88*, 73–86.

Kang, L. (2021). Street architecture landscape design based on Wireless Internet of Things and GIS system. *Microprocessors and Microsystems*, *80*, 103362.

Kannus, K., & Ilvonen, I. (2018). Future prospects of cyber security in manufacturing: Findings from a Delphi study. In *Proceedings of the 51st Hawaii International Conference on System Sciences*. Iran.

Karmakar, A., & Delhi, V. S. K. (2021). Construction 4.0: What we know and where we are headed. *Journal of Information Technology in Construction*, *26*(28), 526–545.

Katiyar, A., & Kumar, P. (2021). A review of internet of things (IoT) in construction industry: Building a better future. *International Journal of Advanced Science Computing and Engineering*, *3*(2), 65–72.

Statsenko, L., Samaraweera, A., Bakhshi, J., & Chileshe, N. (2023). Construction 4.0 technologies and applications: A systematic literature review of trends and potential areas for development. *Construction Innovation*, *23*(5), 961–993.

Waqar, A., Shafiq, N., Othman, I., Alsulamy, S. H., Alshehri, A. M., & Falqi, I. I. (2024). Deterrents to the IoT for smart buildings and infrastructure development: A partial least square modelling approach. *Heliyon*, *10*(10), e31035.

Case Studies on Smart Geotechnical Engineering

13

13.1 INTRODUCTION TO SMART GEOTECHNICAL ENGINEERING

This transition of using smart technologies to geotechnical engineering has brought about revolutionary changes in the way infrastructure management and safety are tackled, all because of an old-school technology known as the Internet of Things (IoT). As such, it is designed to convey a broad range of advancements in applications that improve our understanding of soil and rock behaviour using technologies including real-time data capture analysis. The new connectivity of cars is part of a widespread trend in engineering towards digitalization, related to the development of sensors for monitoring the states of machines and systems and data processing in connection with smart algorithms (Zhang & Lu, 2008).

The IoT is at the core of a change that redefines geotechnical monitoring and analysis. IoT technologies enable a large number of sensors and devices to be deployed in different geotechnical fields across the globe that collect information continuously. These sensors can track the likes of soil moisture, ground vibration, and structural movement with record-breaking accuracy. The data then goes to centralized systems in real-time that can analyse it, creating condition reports or pre-empting issues before they happen.

There are numerous benefits of using IoT in geotechnical projects. Safety, or improved safety, is one of the most basic ways IoT benefits us in our lives, especially with early warnings. A good example of that would be sensors that (a) identify when the soil being built upon is moving or (b) alert engineers to potential shifts in building stability before they become insurmountable problems. The ability to see and anticipate issues before they happen allows for intervention response much sooner, thus reducing the chances of a catastrophic failure, as well as increasing overall project safety (Yusof et al., 2007).

DOI: 10.1201/9781003537625-13

Second, integrating IoT makes us smarter in taking a decision. Engineers ha real-time insights that help them to make data-driven decisions for maintenance or desig tweaks as opposed to relying on incomplete and outdated information. As part of th proactive approach, project management becomes more efficient, and infrastructure tend to last a little longer since problems are addressed before they have gone critical.

In total, the use of smart technologies in geotechnical engineering is changing an bringing with it new instruments for monitoring, analysis, and decision-making. B developing giants, they will change how we manage and maintain geotechnical infra structure—for the better to further evolve.

13.2 CASE STUDY I: IoT APPLICATIONS FOR REAL-TIME LANDSLIDE MONITORING

13.2.1 Project Introduction and Objectives

Over the past few years, landslides have become a major concern for many communitie and infrastructure, especially those in steep terrain or areas with high precipitation. A pilot project was carried out to implement real-time IoT-based landslide monitoring t manage risks associated with it. The main aim of this project is to increase the safet by previous landslide detection, making it possible for early measures to be taken. Th project used advanced IoT sensors and data analytics to continuously monitor landslides thus providing real-time early warning of an impending landslide (Teltayev et al., 2022)

13.2.2 IoT Sensors and Data Acquisition System

The automated landslide monitoring system relies on a network of IoT sensors that monito a series of environmental factors. Principally, these are inclinometers to register soil move ment, piezometers for monitoring groundwater, and accelerometers sensitive to ground vibra tion. Heavy sensors were planted with care points to the slope as well as infrastructure whicl can be used to measure stability and moisture content, respectively. Sensors for human biote lemetry collecting information and data acquisition systems, which transmitted it via telem etry in real-time back to a central monitoring point, were employed (Berglund et al., 2020).

13.2.3 Landslide Prevision of Data Analytics

Using the advanced data analytics tools, the data collected from sensors were processec in real-time. Those tools scoured the data for patterns and anomalies that might be signs of a coming landslide. For instance, the heave of groundwater or ground movements such as earthquake can indicate an increased potential for a landslide. With the help of machine learning (ML) algorithms and statistical models, these systems could accurately forecast possible landslides that were likely to strike soon, affording authorities ample time to take necessary precautions (Babu et al., 1988).

13.2.4 Challenges and Their Solutions in Monitoring

Although using IoT in landslide monitoring has shown to be advantageous, yet many reference a lot of challenges during the solicitation of this project. A significant challenge was getting the sensors to work reliably and last in tough environments. To counteract this, the sensors were installed in weatherproof casings so that regular maintenance could take place. The sensors also produced vast amounts of data, which was another major challenge to be tackled. To address this, more stalwart data processing techniques and cloud-based storage solutions needed to be used for fast management of the data and its sensitive analysis (Morin, 2021).

In the end, this real-time landslide monitoring project proves how the use of IoT can role in boosting our environmental safety infrastructure. The project delivers real-time insights into potential slope stability increasing the detection and response of landslide threats, thereby protecting communities and infrastructure from exposure to a looming threat.

13.2.5 Smart Footings and SHM

13.2.5.1 Monitoring settlement in large buildings

Settlement issues are primarily observed in big structures like bridges and high rises as the relationship between the load of building and the soil is intricate. The smart foundation project was started to monitor and control settlement in these large-scale constructions to overcome the above-mentioned problems. As of 1998, the project was intended to dynamically monitor settlement trends in real-time and document the long-term stability of its foundations. The project used advanced monitoring technologies to identify any trends of irregular settlement and excessive settlement early, which would allow for the corrective action to be taken in a timely manner to prevent potential damage (Sarsembayeva et al., 2022).

13.2.5.2 IoT-based instrumentation foundation health

At the centre of the smart foundation project is an IoT-enabled sensor network with real-time monitoring to constantly assess the foundation's integrity. To assess the displacements, multiple sensors have been mounted such as displacement sensors for vertical and horizontal movements, strain gauges to measure structural strain, and load cells to assess forces acting on the foundation. This network allowed the sensors to continually report on how the foundation was functioning in real-time. This data was then wirelessly transmitted to a central system, where analysis could show signs of trouble in the foundation (Ahmed et al., 2019).

13.2.6 Data-Driven Decision-Making in Foundation Design

Information about the soil properties and its structure is required for laying the foundation which can also be monitored using IoT sensors so as to plan out a regular maintenance process. Engineers could leverage this data and determine the necessary releveling or bracing of caissons needed, with predefined limits for how much additional settlement had started to occur. So, if sensors detected irregular settlement, engineers could inspect whether additional supports were required or the distribution of load needed to be adjusted. Because of the data-driven methodology, the building could now be managed at optimal levels and structural issues were fully avoided, besides increasing the overall lifetime of the building.

The introduction of smart technologies in the monitoring of foundation movement improved tracking and settlement management capabilities, providing insight into the health of the structure. Showcasing how technology can be employed to increase the safety and efficiency of large infrastructure projects through the use of IoT-based instrumentation was coupled with data analytics, ensuring foundations remain stable and secure for the long term.

13.2.7 Internet of Things Applications in Tunnel Boring Machines

The tunnel boring machine (TBM) is the centrepiece of modern tunnel construction, making it possible to excavate tunnels for logistics and utility facilities quickly and efficiently without hindering rail or road traffic. IoT applications have been implemented with the TBM in this case study, improving its operational efficiency and safety. The TBMs were fitted with a host of sensors that could provide real-time information on cutter head performance, rotation per minute, and thrust forces among other parameters. This data was monitored at all times to make sure that the machine worked ideally and that there were no performance outliers. To achieve a snapshot that shows the complete work in progress of the TBM, engineers combined extensive IoT technologies so they could see how well it was working, and make adequate changes to keep productivity up while avoiding future issues (Taha et al., 2000).

13.2.7.1 Monitoring for surface movements and tunnel deformation

One of the key things to consider in tunnels especially when doing tunnelling is ensuring that the soil surrounding does not continue to shift or move during drilling since this can affect the integrity of the tunnel itself. The project tackled this problem by using a combination of IoT sensors to track ground movement and stability in the tunnels themselves going forward. Extensometers, inclinometers, and strain gauges were installed in the tunnel and surrounding area to measure ground displacement, structural strains, and deformations. The collected data made it easier to evaluate the safety of the tunnel, and early signals that are good warning indicators for problems like ground subsidence or deformation could be provided. It could quickly act if required to prevent the danger of collapses due to deformation, thus guaranteeing that the tunnel drive was working safely (Townsend, 1985).

13.2.8 Predictive Maintenance and Risk Management

Predictive maintenance and risk management were some of the key advantages afforded by IoT in tunnelling operations. For example, readings from the TBMs and ground sensors were analysed against real-time data and historical information to detect early warning signs of potential maintenance issues or risk factors. For example, sensors would report that components are worn out sooner than expected/that the ground is moving in a weird way, so maintenance for these items could be planned before things get bad. Such a future-looking thesis, not only ensured the overall efficiency of the TBM but also reduced its downtime while extending the lifecycle of the equipment. It also identified a means to better manage risks by tackling any new threats that arose, thus preventing expensive delays or safety incidents.

13.2.9 Applicability of Wireless Sensor Networks for Slope Stability

Slopes are instrumental in maintaining the stability of a natural reservoir and are essential for the preservation of structures, infrastructure, or human lives (in areas subjected to landslides or erosion). This case study presents the deployment of IoT-enabled solutions for slope stability assessment using wireless sensor networks. Sensors in the networks measured a variety of slope variables, which included soil moisture, ground movement, and pore water pressure introduced into sensors across the slide surface. The wireless technology incited a flexible sensor placement and real-time data transmission, obviating the need for yuppie wiring, thus offering scalability and improved efficiency of the monitoring system. This strategy allowed for an extensive examination of the slope, which gave them more significant meaning to how this part of the slope is (or was) acting (Teltayev & Suppes, 2019).

13.2.10 Data Collection and Analysis in Real-Time

IoT sensors are used to enable continuous, real-time data acquisition from the slope. Information about anything from the behaviour of soil and humidity to settling of structures was detected by the sensors and transmitted wirelessly to a central data processing system. This data was processed through advanced data analytics tools to provide real-time slope stability feedback. The system tracked patterns and anomalies in rea class data—it could see goodness and badness using math quirks. An example alert might be sudden changes in moisture levels or unexpected ground movements that would require prompt action to help prevent slope failure.

13.2.11 IoT Impact on Slope Management and Safety

The inculcation of IoT technologies in slope stability assessment resulted in major enhancement to slope management and safety. Since monitoring also took place in real-time, engineers received early warnings that an application may become unstable which allowed them to take action before it became a major issue. In addition to increasing the safety of local infrastructure and surrounding communities, this also significantly limited the costs associated with emergencies, response, and repair. In addition, continuous data collection made possible historical records that could be used to improve risk assessment models further and enable better planning of future slope stability.

13.2.12 IoT Seismic Monitoring and Ground Vibration

Earthquake-proof structures that can withstand the forces of seismic activity in an earthquake-prone region are vital to keeping property damage and personal injury to a minimum. In a case study on the Adoption of IoT Technologies for Applications in Seismic Monitoring and Ground Vibration Analysis, they used IoT sensors to measure the seismic forces and ground vibrations during earthquakes. This rapid response information with sensors like accelerometers and seismometers reported shaking intensity but also the frequency of seismic waves. Through continuous monitoring of these parameters, engineers would be able to build a detailed picture of how earthquakes impact different areas and structures, informing better design strategies for quake resistance (Daniel et al., 2022).

13.2.13 Smart Respond Earthquake Ecotechnological Engineering

IoT embedded in the earthquake-resistant geotechnical designs achieved smart responses to the seismic event. In the event of an earthquake, IoT systems might automatically control various damping devices in buildings or transmit signals (of the "help me!" variety) according to emergency plans. One example would be sensors that could identify substantial ground movements and then deploy base isolators or energy dissipation devices to reduce the impact of seismic forces. The ability to give real-time responses has enabled them to protect buildings and save lives from damages. Furthermore, data from IoT-enabled monitoring systems could give operators information on how their experience faced with a storm translated to measurable benefits and how they'd be even better next time.

13.2.14 Reinforcement and Retrofitting Techniques Based on Data

Analysis and data collection after the earthquake were of paramount importance to enhance future earthquake-resistant designs. The ground motion data collected by IoT sensors from the earthquake events were analysed for a thorough evaluation of the existing geotechnical design performance and to locate hotspots that required strengthening or retrofit. If the data identified that specific buildings were subjected to excessive stress or movement, then this information could be used by engineers to develop targeted strengthening solutions. By making it possible to gather real-time data on how buildings performed during strong shaking, IoT technology also made this "data level" as important as the seismicity (magnitude and distance) when making decisions regarding reinforcement strategies and retrofitting technologies. The system also led us to develop more resilient new systems and retrofits that raised the performance of prevailing structures (Latiffi et al., 2013).

13.3 CASE STUDY II: SMART DAMS AND EMBANKMENTS—A HOLISTIC SOLUTION

13.3.1 Remote Water Pressure and Seepage Monitoring Using IoT

Dams and levees are crucial for disaster prevention and the ability to monitor their strength could prevent disastrous pranks. Using IoT technologies, a case study was carried out on better monitoring of water pressure and seepage in dams and embankments. These include the installation of smart sensors whereby a number of monitoring elements including pore water pressure, seepage rates, and water levels can be calculated in real-time throughout these structures. This would give an early warning to the engineers based on continuous data, if seepages or increased pressure are threatening the structural integrity. It was this capability of IoT to provide real-time data that secured the integrity of these vital systems.

Real-time data streaming: One of the biggest benefits of IoT-enabled systems is the facility to collect and analyse data in real-time. This meant a monitoring infrastructure that could deliver real-time information about the state of their structural health and stability, especially for dams and embankments. With IoT sensors in place, we the developed advanced data analytics platforms that could correlate the stream of incoming data to create insights and identify patterns or aberrations which implied risk. For instance, any abnormal rises in water pressure or seepage rates could be instantl grounded as a basis for follow-up or intervention. By answering these questions i real-time, folks could take proactive action ensuring no new collapsed dam ever sa the light of day and that maintenance could be planned in a timely way.

13.3.2 IoT for Disaster Preparedness

IoT not only improved regular monitoring but also helped in disaster management fc various dams and embankments. When a structural problem occurs or there is a seriou weather event, IoT systems would be able to alert authorities and emergency teams wit automatic alerts. Such systems were designed to give key alerts on the state of the dan including weak points or impending failure that would provide an accelerated and mor aligned contingency response. What is more, IoT sensors may provide real-time wate levels and pressure data that can allow modelling of potential flood scenarios, enablin for example emergency teams to predict the impact and prepare for evacuation or othe protection actions. This has considerably increased the efficiency of disaster management and enhanced dam safety in general.

13.4 CASE STUDY III: IoT FOR GEO-ENVIRONMENTAL MONITORING

13.4.1 Waste Management and Suppression of Contamination

In the field of civil work (especially in geotechnical engineering), waste and pollution detection are extremely beneficial to environmental preservation. Case study o the use of IoT in monitoring waste disposal sites for immediate contamination risks analysis. Smart sensors were used to measure the site's hazardous material levels—from methane emissions to toxic leachate—so as to ensure no contaminants spreac into the surrounding land or water tables. These IoT sensors have a 24/7 monitoring facility for environmental conditions, which could detect any breaches or irregularities in advanced. Using this system to automatically detect oil allowed for immediate intervention, mitigating the risk of widespread contamination as well as its subsequent ecological impacts (Zhussupbekov et al., 2021).

13.4.2 IoT Solutions for Groundwater Monitoring

Another important area of geo-environmental management is to maintain the quality of groundwater and IoT, the latest revolutionary approach was created for the same. The water sources near larger industrial or waste sites are monitored by IoT-based groundwater monitoring systems from time to time to get the levels of pollutants and their pH level, and temperature scale. Sensors were placed at different depths to monitor the required groundwater conditions. The real-time availability of data from these sensors allowed us to identify changes or trends in the water, which could be indicators for potential contamination. Real-time monitoring system helped them to identify the change in groundwater quality instantly and take preventive actions accordingly so that water could be maintained at a safe level.

13.4.3 Smart Technology-Based Environmental Impact Assessments

Also, IoT mechanized and improved the way we did Environmental Impact Assessments (EIAs) for geotechnical projects. In the past, an EIA would be based on periodic data collection and manual assessment methods which meant that crucial elements of environmental impact could be missed. The difference with IoT is that real-time monitoring of critical environmental characteristics such as air quality, soil composition, and noise levels meant the assessment could be more living and complete. By adopting this data-driven method, geotechnical engineers could keep their eyes on any natural entities that changed over time during project implementation and go ahead accordingly. By embedding IoT within EIAs, any environmental hazards that were identified early on could be addressed in real-time, and project outcomes were both more sustainable and responsible.

13.5 CASE STUDY IV: IoT IN GEOTECHNICAL RISK MANAGEMENT

The geotechnical industry experiences significant risks including the effects of natural hazards, structural collapses, and environmental influences on projects. It used IoT-sensing technologies so as to be able to detect and manage these risks well in advance. Through the use of a network of sensors installed in key sites such as slopes, foundations, and tunnels, IoT systems dynamically measured variables like soil displacement, pressure variations, and stress on the structure. These sensors took real-time data, making it possible for engineers to react quickly when they detected initial signs of instability or wear. By flagging up risks as they presented themselves, geotechnical teams were able to address these in advance of them becoming a bigger problem and realize safety and reliability improvements across their infrastructure.

13.5.1 Proactive and Predictive Making an Effort to Be Prepared before Something Actually Happens; Utilizing Predictive Analytics

One of the main advantages of IoT in geotechnical risk management is utilizing predictive data analytics to help with the implantation of proactive measures. IoT sensors would gather the data, and advanced analytics tools were used to trace historical trends and existing conditions that could lead to failure. For instance, engineers may be able to predict when a building will start shifting and move to shore it up before that happens by examining records of ground sinking or vibration. Using this predictive strategy, not only reduced the probability of having to pay for expensive repair bills but also prevented massive catastrophic disasters such as landslides or foundation failures. With its predictive analysis capabilities, IoT represented quite a departure from the reactive approach to geotechnical risk management, and an opportunity for early interventions being gradually driven by data.

13.5.2 Insurance and Economic Implications of IoT-Driven Risk Mitigation

IoT also has economic and insurance consequences in geotechnical risk management. The adoption of risk mitigation strategies derived from IoT in this vertical has also reduced insurance premiums associated with geotechnical projects through the elimination of infrastructure failures. Insurance companies realized that permanent monitoring plus predictive analytics actually made them less vulnerable and able to provide better terms for projects using IoT technologies. The savings inherent in stopping predictable failures and minimizing downtime justified the costs involved with an investment in IoT for large geotechnical projects. The economic value created by them in conjunction with IoT that could mitigate uncertainties through effective real-time monitoring and alerts impressed on the importance of integrating smart technologies in geotechnical risk management strategies.

13.6 CASE STUDY V: IoT-BASED SMART PAVEMENT SYSTEMS

This application of IoT technology falls under the category of research, it is implored to observe smart pavement system and apparatus for monitoring and maintenance of those pavements. Without regular maintenance and upgrades, pavement deformation problems such as cracking or subsidence can become safety hazards for motorists that may be expensive to repair. The IoT sensors were embedded in the road surface to continuously record temperature fluctuation, load-bearing stress, and movement of pavement. The sensors were able to detect deformation patterns in the pavement early on, so engineers could remotely check a road's condition immediately. By executing this plan, the IoT system was able to ensure interventions were early enough that minor issues such as roadways not being dried up properly could prevent structural failures.

13.6.1 IoT for Roadbed Health and Maintenance Planning

Sensors based on IoT technology were instrumental in examining roadbed health. Subsurface conditions were assessed by sophisticated sensor networks, and parameters such as soil moisture, compaction, and the stability of the underlying layers were monitored in real-time. Such systems enabled accurate maintenance planning, which made it possible to carry out repairs in exactly the location and at exactly the time that needed them. As a result, the roadbed maintenance plan was based on data, which made it possible to find an optimal balance and avoid unnecessary repairs, while not allowing degradation in the conditions of the roadbeds to go unaddressed. This approach to roadbed health maintenance led to cost reductions since maintenance became less frequent and more effective.

13.6.2 Infrastructure Lifespan Extension

The IoT-based pavement monitoring not only helped to manage performance in real-time but also extended the infrastructure's overall lifespan. IoT technologies helped to detect problems at an early stage and perform preventive maintenance based on the analysis of information: this way, wear was minimized and the premature ageing of pavement systems was eliminated. In addition to that, performance analysis helped to develop predictive models that made budgeting for major overhauls more accurate. Since these replacements could be predicted many years in advance, they did not become unexpected expenses: such planning expanded the period between major overhauls and replacements and made the most effective use of resources. Thus, the IoT-enabled pavement systems contributed to longer longevity, safety, and economy of infrastructure (Ham et al., 2016).

13.7 CASE STUDY VI: IoT APPLICATIONS IN OFFSHORE GEOTECHNICAL ENGINEERING

One of the new challenges facing offshore geotechnical engineering is a product of the harsh and dynamic environments in which many marine structures are being built. In this study, IoT devices were used to track marine foundations and platforms (e.g., offshore oil well and wind power), with the objective of preserving their structural integrity. IoT sensors were installed on these structures which enabled continuous monitoring of parameters like the stability of a foundation, distribution of loads across structures, and forces exerted by the environment (wave influences, tidal motions). The capability of monitoring in real-time meant that engineers could assess the status of marine foundations and respond to any developing issues before they became threatening, therefore guaranteeing safety and lifetime for offshore infrastructure.

The sensors used must operate under some of the most extreme conditions encountered in offshore geotechnical engineering. They are built to be able to stand salt water, pressure, and currents; this enables continuous collection of high-accuracy data in these extreme conditions. Even in some of the most isolated and remote areas data, advanced sensor networks provided real-time updates about soil movement, erosion, and seabed stability. It enabled engineers to acquire this data in real-time and gave them better than ever access into the behaviour of their offshore structures, so they too could react faster to changing conditions and make smart decisions on maintenance and safety.

Having covered the general applications of IoT in geotechnical engineering, we would like to give a specific example of how and why offshore geotechnical engineering is utilizing IoT in follow-up articles, exemplified here using monitoring for the foundations of an offshore wind farm. Wind turbines experience considerable stress from wind and ocean currents, so it is important to secure their ballast. Sensors were fitted on the seabed and wind turbine foundations which used IoT technology to record stress levels, vibrations, or movement of marine sediments. This data was then analysed and used to predict the needs for maintenance, reduce risk related to foundation failure, and finally optimize the design of foundations. The integration of IoT in offshore wind farm management has helped its operators to enhance the performance and safety of these renewable energy structures, encouraging a wider recognition of sustainable energy solutions.

Geotechnical engineering is enhanced through predictive analytics powered by the integration of AI/ML capabilities in IoT technologies. The data collected is vast and comes in real-time, as the sensors are collecting the data around-the-clock for all IoT-based geotechnical projects. The data is processed through AI and ML algorithms which detect patterns, forecast possible hazards, or provide early warnings for issues like ground instability or structural failure. Predictive models like these help engineers predict problems before they happen so that they can take proactive action to reduce the risk of expensive and dangerous repairs. AI used in this way allows geotechnical engineers to optimize designs and improve safety and performance for infrastructure projects.

A wide range of geotechnical projects has used AI on input data from IoT devices successfully. For instance, in locations known for landslides, IoT devices detect activity on the ground which is monitored by digital sensors—of ground movement, rain, or soil moisture. AI models analyse this data, forecasting landslides with impressive accuracy to enable evacuation and other mitigation initiatives well in advance. For instance, in a case study relating to TBM, AI is applied for the analysis of sensor data based on ground pressure, vibrations, and material hardness. It is because of this analysis in real-time that TBM operations are being optimized so as not to generate stability problems in the structures surrounding these works, reducing physical accidents. ML algorithms have leveraged the use of AI-augmented IoT solutions to predict settlement rates in larger structures, assisting in foundation designs and preventive maintenance decisions.

A future in which geotechnical engineering mainly relies on AI integrated with IoT will likely see the emergence of several innovations to include autonomous systems, digital twins, and further advances in ML techniques. As of late, virtual models of physical geotechnical systems, known as digital twins, are gaining popularity and engineers can now simulate different situations and foresee how these systems will behave under certain circumstances. Once AI models mature, they can process more intricate geotechnical data and provide better predictive insights and real-time decision-making. Finally, with the help of AI-driven optimization algorithms, designs will also be more efficient to construct and will have less harm on the environment. More so, we will see the edge computing innovations provided for by software-defined architectures to bring in even higher levels of performance on IoT-based systems performing quicker data processing at the site from where the data is collected when these are developed and used onto Dell laptops.

13.8 CHALLENGES AND FUTURE DIRECTIONS IN SMART GEOTECHNICAL ENGINEERING

The potential that can be achieved via the transformative capabilities of IoT and smart technologies within geotechnical engineering is matched by a few challenges. Indeed, the endurance of sensors in severe environments, reliability in data transmission, and requirement for an uninterrupted power source are technical difficulties. From an operational standpoint, it takes the integration of IoT systems into current infrastructure and data flow roadblocks on a regular basis. Second, large-scale deployments are still often difficult as the cost of installing the sensors, maintaining them, and storing data is prohibitively expensive in many situations even if installation costs were low the initial costs to turn on one would negate any financial sense in deploying a large network of LoRA devices particularly in developing nations. Moreover, the shortage of trained personnel to handle and interpret massive amounts of IoT data makes it difficult for anyone to adopt.

Seamless data integration with IoT sensors: Integration of data from various IoT sensors is considered one of the key challenges in smart geotechnical engineering. Sensors are numerous, they generate data in different formats, and bringing all this together in a system for faster than real-time analytics poses an overall integration challenge. To sidestep this complexity, emerging work investigates the advancement of data fusion techniques that enable different types of information to be fused seamlessly into a shared coherent framework. Edge processing is also being implemented for the purpose of processing data close to the sensor, thereby minimizing latency and speeding up decision-making. Better software platforms are also coming along that can operate on datasets at a large scale more effectively and reliably, such as enabling data scientists to read in all the data using the software at once so it is not being pushed all the way down to raw bytes into the engineering layer. This can be extended to standardizing data protocols between different geotechnical projects, further facilitating the easy integration of IoT solutions at scale.

When it comes to IoT implementations for geotechnical projects, scalability is an absolute requirement. Here are a few case studies which we considered telling them successful in nature and know how IoT have been scaled to bigger projects and how the IoT cloud services can be a game changer for your business. For example, IoT sensors have been used to monitor water pressure, seepage, and structural stability of large infrastructure projects such as dams. While growing one of these systems the ability to manage additional data is important. For example, in landslide monitoring projects, IoT networks that began with only a few sensors have migrated to larger areas while keeping the same level of data quality, proving their scalability. Additionally, AI and ML systems are shown to be of help in dealing with the augmented data load, improving predictive models as more and more data points are factored in. After all, they offer a blueprint of how IoT-enabled geotechnical solutions can be scaled, demonstrating that it is possible to bring these reliable capabilities up in large-scale deployments even as the mechanics of scale manage both challenges and system integrity.

The review on a few of the case studies reveals sharp findings and take-home messages in terms of best practices that can help make your IoT seamlessly merge with your geotechnical engineering endeavour. Perhaps the biggest lesson learned here is about early embrace and integration of IoT technologies in projects. Preliminary investigation of the fault-tolerant system, such as scalability, may change forever without damage to existing system performance. The Big Brahma: Real-time data is collected and analysed through sensors. Proactive decision-making can prevent significant landslides, settlements, or structural failures by investing in the IoT for infrastructure (insight 2). The collaborative approach necessary to combine expertise from geotechnical engineers, data scientists, and IoT specialists is critical as well to ensure that the technology not only is a good idea on paper but also can be implemented successfully. Additionally, it is important for successful projects to include sensors installed in a redundant and resilient manner with corresponding interrogation strategies that result in continuous reliable data.

IoT introduces a lot of value to geotechnical engineering by changing how we currently monitor infrastructure and manage the risks. Obviously, the major sector which can gain value here is in safety improvements allowing real-time monitoring to spot risks such as landslides, seismic activity, and structural instability at an early stage, thus saving infrastructure and possibly lives. IoT analytics also makes it easier to make more informed decisions, mitigating the risk of expensive failures, and enables project designs that are more sustainable. Predictive maintenance, meanwhile, takes IoT economic value a step further by promoting the sustainable and adaptive aspects of infrastructure while eradicating unexpected downtimes. In environmentally sensitive projects, IoT technologies also help contribute to sustainability by streamlining the resource management process and minimizing the overall environmental impact of geotechnical works.

On the basis of the lessons learned from the case studies, a few suggestions can be made for smart geotechnical project entities. During the planning phase, project teams should integrate IoT solutions which should be prioritized for easy deployment and to prevent retrofitting issues. The second is a well-crafted data management system, which incorporates not just the acquisition and analysis of data but also ways in which one can store, secure, and allow access to that data. As regards to future trends and needs, the training and development of staff are suggested to be addressed who can operate/utilize the systems seamlessly and accurately due to their combination of specialist geotechnical engineering/IoT technologies. Finally, I encourage greater collaboration between the geotechnical engineers and the IT specialists, and even ecological individuals to create more creative and sustainable IoT-driven solutions. By following these recommendations, you can leverage the benefits of IoT in geotechnical engineering and make your infrastructure projects smarter, safer, and more sustainable.

13.9 CONCLUSION

To wrap up the series on Smart Geotechnical Engineering, it is clear that IoT technologies allow "for a whole new set of possibilities" revolutionizing the field in profound ways. Again, this chapter has run the gamut from real-time monitoring of landslides to using smart technologies in tunnelling operations. Through this combination of IoT sensors and data analytics, we can now see how these technologies allow for the early identification of issues, increase the support of foundations, and streamline maintenance processes. These use cases—real-time landslide monitoring, smart foundations, and earthquake-resistant design—demonstrate the advanced capability of IoT to bring about a paradigm shift in geotechnical practices enhancing safety, efficiency, and decision-making. In this publication, each case study demonstrates the particular applications but also reveals challenges and solutions confronted, thereby providing lessons for future activities. This can be demonstrated by his work developing an IoT-based intelligent slope stability assessment and a smart dam monitoring system that provide real-time solutions to critical problems, thus improving infrastructure resilience. It also helps management to have the predictive analysis and modern risk management using AI and ML. Looking to the future, geotechnical smart engineering will mean ongoing improvement—new technologies are just beginning to be adopted that allow for even more advanced and powerful solutions. But needing high-standard technical and operational standards will be necessary for the effective deployment of IoT. When combined with suitable regulations and practices to overcome key challenges, the geotechnical engineering industry remains well-positioned to take full advantage of the smart technologies that continue to develop.

REFERENCES

Ahmed, I., Xu, D., Rouphail, N., & Karr, A. (2019). Lane change rates at freeway weaving sites: Trends in HCM6 and from NGSIM trajectories. *Transportation Research Record, 2673*(5), 627–636.

Babu, T., Jose, A., Sridharan, A., & Abraham, B. M. (1988). A study of geotechnical properties of Cochin marine clays. *Marine Geotechnology, 7*(1), 189–209.

Berglund, E. Z., Monroe, J. G., Ahmed, I., Noghabaei, M., Do, J., Pesantez, J. E., & Levis, J. (2020). Smart infrastructure: A vision for the role of the civil engineering profession in smart cities. *Journal of Infrastructure Systems, 26*(2), 03120001.

Daniel, I., Pesantez, J., Letzgus, S., Khaksar Fasaee, M. A., Alghamdi, F., Berglund, E., & Cominola, A. (2022). A sequential pressure-based algorithm for data-driven leakage identification and model-based localization in water distribution networks. *Journal of Water Resources Planning and Management, 148*(6), 04022025.

Ham, Y., Han, K. K., Lin, J. J., & Golparvar-Fard, M. (2016). Visual monitoring of civil infrastructure systems via camera-equipped Unmanned Aerial Vehicles (UAVs): A review of related works. *Visualization in Engineering, 4*(1). https://doi.org/10.1186/s40327-015-0029-z.

Latiffi, A. A., Mohd, S., Kasim, N., & Fathi, M. S. (2013). Building information modeling (BIM) application in Malaysian construction industry. *International Journal of Construction Engineering and Management, 2*(1), 1–6. doi: 10.5923/s.ijcem.201309.01.

Morin, G. (2021). What BIM can add to geotechnical engineering. *BIM+*. https://www.bimplus.co.uk/what-bim-can-add-geotechnical-engineering/.

Sarsembayeva, A., Zhussupbekov, A., & Collins, P. E. F. (2022). Heat and mass transfer by vapour in freezing soils. *Energies, 15*(4), 1515. https://doi.org/10.3390/en15041515.

Taha, M. R., Hossain, M. K., & Mofiz, S. A. (2000). Behaviour and modeling of granite residual soil in direct shear test. *Journal of the Institution of Engineers, Malaysia, 61*(1), 27–40.

Teltayev, B., Oliviero Rossi, C., Aitbayev, K., Suppes, E., Yelshibayev, A., & Nugmanova, A. (2022). Freezing and thawing processes of highways in Kazakhstan. *Applied Sciences, 12*(23), 11938. https://doi.org/10.3390/app122311938.

Teltayev, B. B., & Suppes, E. A. (2019). Temperature in pavement and subgrade and its effect on moisture. *Case Studies in Thermal Engineering, 13*, 100363. https://doi.org/10.1016/j.csite.2018.11.014.

Townsend, F. C. (1985). Geotechnical characteristics of residual soils. *Journal of Geotechnical Engineering, 111*(1), 77–94. https://ascelibrary.org/doi/10.1061/%28ASCE%290733-9410%281985%29111%3A1%2877%29.

Yusof, M. A., Mohammad, N., & Mat Derus, Z. (2007). Excusable and compensable delays in the construction of building projects: A study in the states of Selangor and Wilayah Persekutuan Kuala Lumpur, Malaysia. *Journal of the Institution of Engineers, Malaysia, 68*(1), 21–26.

Zhang, Y., & Lu, L. W. (2008). Introducing smart structures technology into civil engineering curriculum: Education development at Lehigh University. *Journal of Professional Issues in Engineering Education and Practice, 134*(1), 41–48.

Zhussupbekov, A., Alibekova, N., Akhazhanov, S., & Sarsembayeva, A. (2021). Development of a unified geo-technical database and data processing on the example of Nur-Sultan City. *Applied Sciences, 11*(1), 306. https://doi.org/10.3390/app11010306.

Index

For Product Safety Concerns and Information please contact our EU representative GPSR@taylorandfrancis.com Taylor & Francis Verlag GmbH, Kaufingerstraße 24, 80331 München, Germany